CHAOS ★ FROM ★ THE ★ SKY

★ The number of reported U.S. tornadoes is skyrocketing. The average rate nears 1,200 a year, *six times* the rate in 1950.

★ Twisters are increasingly reported in states where residents assume they "never" happen, such as California. The Los Angeles basin alone has been hit by more than 80 tornadoes since 1962.

★ Tornado funnels grow up to a mile wide and can destroy an entire town in seconds. The worst in our history killed 700 people in a few hours.

★ "Tornado chasing" has become one of America's wildest amateur sports. For a steep fee, you can hire a guide to take you on a tornado chase.

★ History's first "tornado chaser" was Benjamin Franklin, who pursued a small whirlwind on his horse, snapping his whip at the storm.

★ Scientists have speculated about "killing" tornadoes by launching monstrous balloons into the funnels, by dropping chemicals and metal wires into thunderclouds in hopes of "short circuiting" the storms, and by building protective barriers around cities.

★ A tornado almost changed the history of rock music. One narrowly missed young Elvis Presley's home in Tupelo, Mississippi, in 1936, then went on to kill more than 200 people and injure 700.

★ Dangerous myths about tornadoes: you should open windows as a twister approaches, hills and valleys will protect you against a tornado, and your TV will glow ominously as one approaches, giving you ample warning.

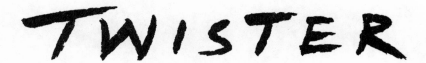

TWISTER

THE SCIENCE OF TORNADOES AND THE MAKING OF AN ADVENTURE MOVIE

KEAY DAVIDSON

POCKET BOOKS

New York   London   Toronto   Sydney   Tokyo   Singapore

WORLDWIDE PUBLISHING™

An *Original* Publication of POCKET BOOKS

POCKET BOOKS, a division of Simon & Schuster Inc.
1230 Avenue of the Americas, New York, NY 10020

TWISTER copyright © 1996 Warner Bros. and Universal City Studios, Inc.
Copyright © 1996 by Keay Davidson
Photos from the motion picture by David James and Ron Batzdorff copyright
© 1996 by Warner Bros. and Universal City Studios, Inc.
Back cover photo © 1989 by Roy L. Britt/Weatherstock

ISBN: 0-671-00029-2

First Pocket Books trade paperback printing June 1996

10 9 8 7 6 5 4 3 2 1

POCKET and colophon are registered trademarks of Simon & Schuster Inc.

Text design by Stanley S. Drate/Folio Graphics Co., Inc.

Printed in the U.S.A.

*This book is dedicated to the memory
of Dr. Arnold Shankman,
a historian, teacher, and friend
who departed this life much too soon.*

ACKNOWLEDGMENTS

First, thanks to my family, particularly my sister Georgia "Tina" Davidson, a fellow tornado buff. She gave me the idea for this book. Second, to Phyllis Heller, who encouraged me to stick by the idea when it was lost at sea, and who steered it toward the right agent, Jane Cushman. And third, to my boss Phil Bronstein at the *San Francisco Examiner*, who generously granted me a sabbatical to explore the bizarre world of tornadoes.

Special thanks to Dick Rogers, brainstormer extraordinaire; everyone at the National Severe Storms Laboratory, especially the amiable and admirable Erik Rasmussen; Dane Konop of the National Oceanic and Atmospheric Administration; Dan Pendick of *Earth Magazine;* Rick Gore, my editor at *National Geographic;* the operators of Toronto Science Island, where I discovered meteorology at age ten; and the weather writers who influenced my youth—Paul E. Lehr, R. Will Burnett, Herbert S. Zim, Philip D. Thompson, and Robert O'Brien. Also, a belated thanks to five people who made a difference, each at a crucial moment in my life, but who probably don't know it: Robert Morgan, Jim Squires, June Smith, Will Hearst III, and Paul Freiberger.

Naturally, I alone am responsible for any mistakes herein.

C ONTENTS

CONTENTS

F OREWORD

What *Jaws* did for sharks, the film *Twister* is likely to do for
tornadoes. The classiest movie "thrillers" do not merely thrill;
they also educate and inspire the curious to learn more about
the natural world around them. Two decades ago, *Jaws* encouraged untold flocks of youngsters to head for libraries or
bookstores and ask clerks, "Do you have any books about
sharks?" Likewise, *Twister* may stimulate young and old to
learn more about tornadoes, the most astounding and fright-
ening of atmospheric phenomena. Oddly, no major publisher
has issued a popular nonfiction book devoted entirely to tornadoes in many years. This book aims to fill that void. Hope-
fully, the book will inspire readers to a wider interest in
weather, atmospheric science, and earth science in general.
Such interests are especially worth cultivating at a time when
the planetary ecosystem is endangered. Earth requires careful
cultivation, and the best cultivators are scientifically savvy citizens.

Also fascinating is the making of *Twister* itself. Every
filmmaker who portrays a natural disaster—a tornado or a
hurricane or an earthquake or a tsunami or a volcanic erup-
tion or a shark attack, or whatever—faces the same basic
questions: How can this natural phenomenon, so alien to the
ordinary filmgoer's experience, be rendered "realistically" on
the screen? How can the realistic look be balanced with the
audience's expectations? (For example, some filmgoers may
assume that tornadoes resemble the snaky specter of *The Wizard of Oz*, a specter that was, in fact, closer to folklore than
reality.) How can the scientists who study the phenomenon
be portrayed honestly without stooping to typical Hollywood

caricatures of "nerds"? And how can the film's realism be "charged" for dramatic purposes without sacrificing scientific credibility?

In short, how can the silver screen do justice to the marvels of nature and, thereby, covertly educate moviegoers while overtly entertaining them? It is a daunting task, which can be solved only by an army of talented and passionate artists—from actors to editors and sound engineers to computer experts.

Very few people have gotten a good, close look at a tornado—say, from a distance equivalent to a few city blocks. And almost no one has been *inside* a tornado as it rages overhead. But any filmgoer can now undergo that unparalleled experience, thanks to the recent marriage of high-speed computers and Hollywood filmmakers. *Twister* does for tornadoes what *Jurassic Park* did for dinosaurs: It brings them to life on the screen, in computer-generated images that are virtually indistinguishable from photographic reality. Consequently, every filmgoer has the potential to become a tornado "chaser"—and without risking her or his life in the process.

Twister dramatizes the work of one of the strangest elites in science: tornado chasers. Hollywood has a long history of portraying scientists as monsters or madmen, rarely as human beings; in the hands of lesser filmmakers, tornado chasers would probably be depicted as suicidal lunatics. In contrast, *Twister* portrays them as humans with recognizable feelings who are driven by genuine curiosity about nature. *Twister* director Jan De Bont—who directed the bomb-on-a-bus thriller *Speed*—knew from the start that he *didn't* want lead actors that audiences wouldn't "buy" in these roles. Rather, he wanted actors that the audience could relate to, who not only look like real people but who *act* like them.

That's why Bill Paxton and Helen Hunt struck De Bont as perfect for the lead roles. "From the moment I read the script, I knew whom I wanted: Bill Paxton and Helen Hunt," De Bont says. "It's just intuition; I didn't think about it for a second further." In *Twister*, Hunt and Paxton portray people who are essentially like us, with one glaring difference: They chase tornadoes.

One of Hollywood's emerging major stars, film veteran Paxton won international attention in director Ron Howard's *Apollo 13* as Fred Haise, the symbol of Everyman, the antithesis of macho "Right Stuff" mythology, aboard the stricken spaceship. Audiences instantly empathized with his character, a brave but shivering, sickly and occasionally defensive astronaut—an ordinary-seeming American embarked on an extraordinary adventure.

In *Twister*, Paxton plays a tornado expert torn between loves—love of chasing, love for his soon-to-be-ex-wife, love for his new fiancée (played by Jami Gertz), and anger at the film's "heavy," a former colleague (played by Cary Elwes) whom Paxton accuses of stealing his idea for ejecting sensors into the tornado funnel. Paxton "looks like he's from the country-side. He's born in that region, anyway," De Bont said. "He doesn't have that 'macho' actor's stuff that a lot of stars have going for them, which is so distracting from reality. . . . I like actors to *be* what they have to represent."

De Bont's prescription also fits Hunt. She is the female star of the acclaimed TV comedy *Mad About You*, where she portrays an intense but appealingly funny New York woman married to Paul Reiser. Her brief appearance as David Caruso's wife in *Kiss of Death* gave a warm, poignant core to that otherwise hard-boiled crime film. "She looks like a real person, like your neighbor. And it was really hard to get her. She had a commitment to the TV series; the studio was afraid she wouldn't finish in time to get her back to the TV series, and what were we going to do if we ran over schedule? . . . We finally had to promise the studio an 'end date' to get her back to NBC. We just barely made it."

In *Twister*, Hunt plays a self-willed woman, named Jo Harding, who is driven by a childhood encounter with a tornado to chase and study these bizarre phenomena. She is so driven, in fact, that her husband, Bill, leaves her for Melissa Reeves (Gertz), a reproductive therapist. Reeves has no idea how dangerous tornado chasing is until Bill, while trying to persuade Jo to sign their divorce papers, gets caught up in another tornado chase. The chase involves a major storm that generates numerous tornadoes over a day and a half. During the chase, Jo and Bill's collapsed marriage looms like a ghost, but their continuing banter keeps the script lively in between horrific encounters with tornadoes.

Twister is subtly but unmistakably feminist. Hunt plays the most dynamic female scientist on film in a long time, perhaps ever. Traditionally Hollywood portrayed science as a male pursuit, particularly the physical sciences (such as meteorology, the science of weather). Science-fiction films used women mainly as "love interest" or portrayed them as dizzy females whose primary role was to scream and run away from monsters. There were a few early exceptions, such as Faith Domergue's portrayal of a no-nonsense scientist in Ray Harryhausen's monster classic *It Came from Beneath the Sea* (1955). Not until the late 1970s and 1980s did a striking number of major science-fiction films begin to feature tough women, such as Sigourney Weaver in *Alien* and its sequels and Linda Hamilton in *Terminator* and *Terminator 2*. Even then, action films that showed women actually functioning as *scientists*—doing research, operating computers, etc.—have been almost nonexistent. (A rare recent exception is the Laura Dern character in *Jurassic Park*.)

For years, teachers and education experts have complained that it's hard to persuade young girls to pursue scientific careers partly because they see so few "role models" on TV and in films. Considering its huge potential appeal to young moviegoers, *Twister* is a step toward altering sexist stereotypes about science and adventure. *Twister* "is a movie that has a very strong female role—a woman [Hunt] who is totally in control of herself and her own life," says De Bont (who directed another dynamic female character in his last film—Sandra Bullock, who drove the bus in *Speed)*. "She's not dependent on somebody else, and she knows exactly what she wants to do: She wants to chase tornadoes. She wants to create a better warning system for tornadoes so fewer people get hurt. And nobody can stop her. And that's what I like about her.

"A lot of movies, especially in this adventure genre, tend to make the male part so much bigger and more important than the actress's. It's quite different in this movie . . . I *like* strong women very much. I cannot deal with 'cutesy' girls. I want women who know what they *want*. . . . I think it's really important, because I've worked on a lot of movies where it was not like that, and it really bothered me so much—those female parts were almost neglected, so neglected that if the woman had been left out of the movie nobody would have missed her."

De Bont shot *Twister* on location in Oklahoma and Iowa, where tornadoes frequently tear across the prairie. Many film crews on location endure bad weather, but *Twister*'s makers were wracked by both bad and *good* weather. Bad weather often caused their heavy equipment to bog down in mud on narrow country roads where it was difficult to turn around. Good weather frequently denied them the storm footage vital for the film.

Twister was "probably five times as difficult as *Speed*," De Bont said. "To be honest, if I had known it was going to be this difficult, I might have thought a second time about doing it.

"It was *so* complex. The equipment we needed was so big, and to be on all these rural roads where you could not turn cars around, and we got stuck. . . . We had terrible weather, too. There was a lot of flooding in Oklahoma. . . . We got stuck in the mud, all the equipment and tractors and bulldozers and cranes, and I thought, 'Oh my God, how do I ever get out of here?' There were hailstorms and thunderstorms *continuously.*

"The noise of all those jet engines and wind machines became a major headache. I lost my voice so many times trying to shout over the sound level of the wind machines, I was forced to use hand signals. For the actors it was hard, too, because when you have those wind machines and jet engines aimed at you, you don't act, you just try to survive! Debris flew in their mouths as they were saying their lines, and they'd gag and have to stop talking."

De Bont is no stranger to rugged filmmaking. Born in Holland in 1943, he worked as a director of photography, gaining attention for his work with Dutch directors including Paul Verhoeven. In the United States, he worked on complex, energetic films such as *Die Hard* and *The Hunt for Red October*. In terms of popular renown, his breakthrough film was *Speed*, which introduced a truly mass audience to his flair for threading humor and personalities through complex, high-tech plots.

Despite De Bont's past experience, the pressure involved in making *Twister* was "absolutely horrifying," he says. "The movie was so complex because of all the physical and computer effects. The physical effects on the set were unbelievable. Every day, there're hundreds of people asking you a million questions that will never stop. Our days would start at six in the morning, and my day wouldn't end until ten or eleven in the evening. And we'd be shooting seven days a week."

De Bont hired a meteorologist, Vince Miller, to work with the film crew to alert them to coming weather conditions. De Bont was especially concerned about Midwestern lightning, which is nothing short of spectacular.

"Those bolts were so powerful that when they flashed, it was like daylight—I had never seen anything like that. It was just unbelievable, absolutely beautiful," says Ian Bryce, another *Twister* producer. Meanwhile, "you've got 200 to 250 people out there, and you're responsible for their lives, and there's no basement you can put all 250 people in.

"This is the hardest movie I've worked on," Bryce affirmed, "and I've worked on a *lot* of [special] effects movies."

The filmmakers simulated a hailstorm by moving seven trucks full of ice down from Milwaukee. The ice had been injected with milk to make it resemble hail. Then crew members shoved 400-pound ice blocks into a chipping machine. The ice fragments were sprayed over the actors at a site they dubbed "Hailstorm Hill." While the hail spewed overhead, the crew aimed a 707 jet aircraft engine (mounted on the back of a 48-foot flatbed truck) at the actors. Other crew members tossed branches and other debris into the jet's winds to simulate hurtling tornado debris. It was quite a spectacle.

Producer Kathleen Kennedy said: "We were moving a crew of 250 to 300 people constantly, sometimes three, four, five moves a day. That meant the transportation crews would be up all night long moving trucks, moving base camps, making sure everybody had a place to live during the day.

"It got very, very hot—the Midwest can get 100 percent humidity, 90 degrees. It's pretty rough to be out working in those conditions, day in and day out," Kennedy said.

But it was worth it, added Kennedy, a veteran producer of some of Hollywood's most successful films. "If you want a spiritual experience, you should go spend April to June in the Midwest, because you have never *seen* cloud formations like this! You watch everything in the sky happening in front of you as if you were watching time-lapse photography. We would literally watch cloud towers shoot into the sky and within fifteen minutes one little cloud would rise to become one 30,000 feet high."

John Frazier was in charge of special physical effects. He works in Sunland, California, in a building that, from the outside, looks more like a large auto repair shop than a film fantasy factory. The only external hint that odd stuff goes on within is a large, partly dismantled green dragon in the driveway. (He built the dragon for a ceremony at his daughter's high school.)

Frazier is an easygoing Hollywood veteran who has

worked on numerous films as diverse as *Hoffa*, *Speed*, *Apollo 13*, and Jim Carrey's forthcoming *Cable Guy*. He holds out a box full of what appear to be little spherical gizmos with translucent covers, chromium bottoms, black wires on the outside, and blinking red and green lights on the interior. "We made thousands of these things. Take some." In *Twister*, the balls were scientific "sensors" that the researchers launch into a tornado.

A movie's physical props, like its digital effects, may contain inside jokes that the audience never sees. The sensors were made by a physical effects worker whose nickname is "Fluffy" ("because he looks like marshmallow," Frazier explains). Frazier holds up the translucent sphere and points to the name of an imaginary company neatly printed inside the sphere: "Flufftronics."

Frazier, along with production designer Joe Nemec, also supervised crews that used bulldozers to tear down a section of an Oklahoma town covering two city blocks. The movie studio purchased all homes and buildings in that area for the film, then ripped them apart to portray the aftermath of a tornado.

Twister's trickiest and riskiest physical effect involved a tractor-trailer that is picked up by a tornado, then dropped in a fiery explosion on the highway. Frazier's crew had to strip down the tractor-trailer until it was light enough to be lifted by a crane. Then two stunt experts, representing Jo and Bill, drove in a Dodge pickup truck toward the tractor-trailer as it dangled 75 feet above the ground. The tractor-trailer was wired with explosives. At a precisely specified moment, Frazier pressed a button to release the tractor-trailer. If his calculations were right, the tractor-trailer would hit the ground about 50 feet in front of the approaching stunt crew. A moment's miscalculation could kill two people. The blast came off without a hitch. That one shot cost at least $100,000.

"You'll wake up at all hours of the night, thinking about how to do a particular effect," Frazier says. He also worries about how to pull off the effect without letting the audience know how it was done. Typically, he has nothing to fear; an audience is usually too dazzled by a physical effect to notice the moviemaker's sleight-of-hand. At the end of *Speed*, Frazier says, "you can see the cable towing the bus into the airplane. But nobody has ever seen that cable but me."

*T*wister's setting is as grandiose as its subject: the Midwest, a terrain as rich in myth for Americans as the Aegean is for Greeks. Such a vast setting calls for the right "eye"—the eye of a skilled director of photography.

What makes the Midwestern sky "so interesting is that the terrain is so *flat*—more than half of what you're seeing is sky! So you tend to pay a lot of attention to it," said director of photography Jack Green. "They've got these incredible cloud patterns passing through—clouds that contrast against a clear, intense blue and nearly unpolluted sky."

Green has been in the film business more than forty years. He has worked with major figures such as Clint Eastwood, for whom Green shot (among other films) *Unforgiven*. For *Twister*'s film stock, Green used "very fine-grain film" to make it easy for digital effects artists to add realistic-looking computerized images of tornadoes. (This book's "Afterword" tells about how the digital effects were done.)

Looking back on his *Twister* experience, Green recalls his and the rest of the film crew's seemingly endless battles with bad Midwestern weather—thunderstorms, lightning, muddy roads. "I think the credit list after this show will be as long as the show," he jokes.

*T*ornadoes are terrifying partly because they sound so horrible—"like a thousand freight trains," according to a common description. Consequently, one of *Twister*'s most important artists was the sound effects editor, Steve Flick.

As a painter works with paints to make a palette of color, a sound effects expert works with different sounds to create an "emotional palette," Flick said. A simple example is a creaking door in a haunted house. You "know" what that sound is like; perhaps you can "hear" it in your mind as you read these words. And if effectively used in a film, that sound produces an emotional reaction (suspense or fear).

For *Twister*, Flick and sound designer John Pospisil searched for sounds with a quality that Flick calls "aggressive strangeness." They listened to recordings of sounds ranging from freight trains and roller-coasters to roaring lions.

To make new and different wind sounds, they constructed a box filled with chicken wire, stuck a microphone inside, and placed it on top of a car. Then they rolled the car downhill—turning the engine off so that it wouldn't interfere with the sound recording.

They also reviewed recordings of camels and noted that these creatures emit sounds that are "wet and lugubrious and nasty—they have a tremendous amount of 'globble' sound to them." As he listened to the camel recordings over and over, Flick turned down the pitch, and the camels' sounds developed a moaning, "cavernous" quality that, he felt, nicely captured the eerie vastness of a tornado.

Perhaps no film art is simultaneously as subtle and as powerful as editing. The arrangement (and rearrangement) of pieces of film can make the difference between a mediocre scene and a great one. For example, artful cutting made the shower scene in *Psycho* a chilling landmark in film history.

The editor on *Twister* is Michael Kahn, one of Hollywood's most distinguished film editors. He began his career working on TV editing for *I Love Lucy* and other shows. He won an Academy Award for his editing on *Raiders of the Lost Ark*, and also worked on films as diverse as *Close Encounters of the Third Kind* and *The Color Purple*. *Twister* was his first opportunity to work with digital editing, in which the entire film is digitized, then stored on a computerized editing machine called Lightworks. The device is bigger than a desk and surrounded by computer disk storage equipment in Kahn's Santa Monica office. He sits in front of the computer screen and watches a digitized copy of the movie unfold on the screen, complete with sound. By manipulating dials, he rearranges frames of the film and does fade-ins, fade-outs, jump cuts, and all the other gimmicks of the film editor's trade—without touching a frame of movie celluloid.

The latter work is reserved for Kahn's assistants. After he finishes editing the digitized version of the film, he has the computer print out a list indicating exactly how he rearranged it, frame by frame. He gives the list to an aide, who goes to the film vault, removes the necessary copies of the original celluloid footage, and cuts and rearranges it on a traditional editing machine. For Kahn, Lightworks allows him to be as imaginative and clever as he desires, while others handle the laborious "cutting and snipping." The main benefit of Lightworks, he says, is that it's so easy and comfortable for him to use. "The only problem," he adds with a grin, "was that I gained fifteen pounds—I'm just sitting and pushing buttons."

Throughout the editing of *Twister*, Kahn regularly consulted with De Bont and frequently visited the shooting locations in Oklahoma and Iowa. Film school textbooks are full of "rules" on how to edit a movie, but Kahn seems blissfully antitheoretical. "We used to have more rules than you can shake a stick at. . . . But today I don't think there are any rules. You just do what works . . . what 'feels' right."

And he feels that his editing of *Twister* will enhance exactly the emotional effects that De Bont seeks: Thrills. Curiosity. Suspense. Romance. Terror. *Lots* of terror.

CHAPTER 1

Ground Zero

In Herman Melville's *Moby Dick,* Captain Ahab chased the white whale across the sea. Ultimately he caught it, killed it, and sank with its corpse to an oceanic grave. To others, Ahab seemed mad; to himself, he was a visionary.

Some of today's visionaries (or nuts, depending on one's attitude) chase a grander and even more lethal target—a leviathan of the sky, not of the sea. Envision a dark, cylindrical object several times as tall as the Empire State Building, and many times as wide. It plummets from a thundercloud, then screeches and spins across the countryside, with winds far faster than a hurricane's. It buzz-saws homes in half, turns towns into toothpicks, and scatters cattle and people like so much confetti.

When a tornado nears, most sane people flee. But "tornado chasers" grab their video cameras, board their cars and vans, and race into the countryside, hoping to get as close as possible to the most brutal and baffling of atmospheric phenomena. Most chasers are amateur weather buffs. But a small percentage are scientists, such as the brave souls of the VOR-TEX tornado chasing project. Their goal is to understand how tornadoes form and how better to predict them. Better tornado forecasts will be more and more crucial in future decades, for over the next century or two, as the United States population grows and spreads across the countryside, tornadoes may become the most common form of severe natural disaster.

The United States is already the world capital of tornadoes. The nation whose Olympian mountain ranges and surreal deserts stunned the pioneers is also haunted by the most Olympian and surreal atmospheric event, the tornado—the *Tyrannosaurus rex* of weather. Tornadoes strike many other countries, but nowhere as ferociously or abundantly as they strike the United States—and not only the Midwest; they even hit Alaska! In 1994, tornadoes killed 69 Americans—the highest yearly tornado death rate since 1984—and injured 1,139. Total damage approached half a billion dollars.

Every spring, a battle ensues in the sky over the Midwestern United States. Each day the sun rises a little higher, and as it does it heats the waters of the Gulf of Mexico. Those waters evaporate and creep north into the United States. Somewhere over the plains states, this tongue of warm, humid air encounters cold, dry air left over from winter. Placing warm, moist air beneath cold, dry air is akin to dropping a match into a gas tank: The moisture rises and forms clouds, some of them thunderheads twice the height of Mount Everest. As energetic as many atomic bombs, the ensuing thunderstorms bring lightning, heavy rain, golfball-sized hail, and tornadoes to the American heartland.

Since 1953, twisters have killed more than 3,700 Americans, injured tens of thousands more, and wreaked billions of dollars in damage. Their fastest winds approach 300 miles per hour. (By contrast, a hurricane is defined as a storm with winds faster than 74 mph.) A typical tornado is hundreds of feet wide, but some have been wider than a mile. Tornadoes turn bricks, pianos, and railway cars into missiles. They shoot straws into trees, playing cards into metal, wood splinters into steel. For days after a twister, victims may pluck sand, grass, and slivers of glass from their skin.

Some twister-wracked towns resemble Hiroshima or Nagasaki after the atomic bombs fell. The reporter Ron Seely crept through the dark ruins of Barneveld, Wisconsin, before dawn on June 8, 1984. A tornado had just crushed nine-tenths of the village. Bleeding residents in their nightclothes staggered through the streets, which sparkled with broken glass. Their bodies were covered with powdered plaster from the exploded walls of their homes; their eyes were stunned and vacant, their mouths half open as if ready to cry out. The ghostly plaster coating on their skin was "eerie . . . you could see tears coming down through the plaster on their faces."

Tornadoes often strike suddenly and without warning. Many decades ago, hurricanes struck without warning, too. But nowadays—when weather satellites swarm the sky, coastlines bristle with radar, and TV networks cover hurricanes as they covered the Gulf War, live, with commercials—hurricanes have lost much of their mystery. The aviation age has demystified the heavens and made the sky humanity's backyard. But the sky retains one primal terror, one elusive and ominous yeti: the tornado.

For more than a century, scientists have struggled to understand how tornadoes form. They have learned much, and today's tornado forecasts are the best ever. But we still have much to discover. The "shock troops" of tornado science are a small band of gutsy researchers who, since the early 1970s, have chased hundreds of twisters across the Great Plains. Some have driven (deliberately or accidentally) within a hundred feet of tornado funnels. Their hearts pounded as they filmed or videotaped the mad whirlwinds. Incredibly, no funnel has killed a scientist—yet.

"There're a lot of people who talk big about tornado chasing and then, when they see a tornado, they chicken out," says one tornado chaser, Professor Jerry Straka. "Look, I don't think people who haven't seen a tornado realize, really, how *big* they are. Even a small tornado is *very* big. . . . When you start seeing big chunks of debris flying around, you have second thoughts."

The grandest tornado chase in the history of science was VORTEX, or Verification of the Origins of Rotation in Tornadoes Experiment. It lasted two years and involved about 120 scientists, students, and aides from 14 universities and other research institutions. Clad in blue jeans and sunglasses, they pursued tornadoes across the central Midwest from April to June, both in 1994 and 1995. In theory, they studied each tornado as an ethologist studies an ill-tempered bull elephant— from a safe distance. "If the team leader deems a particular mission to be too dangerous to perform, the mission should be abandoned," a VORTEX memo warned. However, storm chasers' definition of "dangerous" varies from individual to individual. Enraptured by the howling vortex, few fought the temptation to drive a little closer, a little deeper into the maelstrom of flying telephone poles and bouncing tractor-trailers.

Ironically, VORTEX almost flopped. For mysterious climatic reasons, Mother Nature created few exciting tornadoes in 1994 and 1995. But in June 1995, days before the project's demise, the scientists hurried to the west Texas plains to witness a tornado of mythic grandeur. They circled it like flies, poking and probing its tumult with radar beams and video cameras. Thanks to what they saw and learned, tornado research may never be the same.

The VORTEX field coordinator was Erik N. Rasmussen. He is a skinny, friendly, straight-arrow type with short dark hair and small, intense blue eyes. He looks young enough to ask your high school daughter on a date. By all appearances, he's the kind of guy you'd like to join on a tornado chase: amiable but serious when necessary—someone who won't forget to bring the road maps and peanut-butter-and-jelly sandwiches, and who won't suddenly freak and wail "Mother!" if a twister plunges toward your car.

Like many VORTEX scientists, Rasmussen began chasing twisters as a teenager. As a meteorology graduate student at Texas Tech in the early 1980s, his weather forecasts amazed everyone. Some forecasters are incredibly accurate, and no one is sure why; they just *are*. Despite its multibillion-dollar tools—satellites, radar, supercomputers, etc.—meteorology remains something of an art (as you perhaps realized when rain ruined your picnic or spoiled the office softball game). Anyway, young Rasmussen was so prophetic that he became known as "the Dryline Kid," in reference to an ever-shifting atmospheric boundary, the dryline, that separates dry air from moist air and often spawns severe storms, including tornadoes. The magazine *Weatherwise* cited Rasmussen's "renowned . . . ability to pick the exact spot a tornado will explode to [the] ground." That sounds like an exaggeration, but it was literally true on at least one shocking occasion. A video camera recorded the incident.

May 1981: It's an exciting week in Tornado Alley; numerous twisters are breaking out all over Oklahoma. A radar unit scans a tornadic storm near the town of Binger in unprecedented detail, revealing its internal structure as an x-ray reveals a patient's skeleton. The colorful scientist Stirling Colgate risked death by flying an airplane around a tornado and firing instrumented rockets at it.

Meanwhile, east of Oklahoma City, the twenty-three-year-old Rasmussen and his grad school buddies are driving around the countryside, searching for a twister, any twister. They're carrying a sound-recording device that they plan to drop near the funnel. The device will record the tornado's screeching sound. Scientists hope to learn whether a tornado has a specific acoustic "signature." With luck, the sound measurements could lead to a "tornado alarm"—a sound sensor that rings when it "hears" a twister coming.

The black and white videotape shows what happens next: Driving conditions are rotten on the two-lane road. The windshield streams with rain. Gales buffet the car. "We're heading right into the mesocyclone," Rasmussen declares on the videotape sound track. A mesocyclone is a miles-wide, vertical column of rotating air (like a spinning cylinder) within the thundercloud. The mesocyclone may sprout a much thinner, faster column that descends to the ground—the tornado.

The headlights of another car approach from the opposite direction. Moment by moment, Rasmussen records the fast-changing weather: "Rain, no hail." A bolt of lightning sparks off the road to the left. "CG!" shouts a voice in the car. "CG" means cloud-to-ground lightning. Lightning is the biggest danger, next to the tornado itself. In the last quarter of a century, a few chasers have been "singed"; considering their bravado, it's amazing that none have been fried.

"Very strong winds, rain and hail, moving very rapidly eastward just ahead of us. Watch for possible debris. We're entering a high danger zone." Rasmussen sounds as cool as Captain Kirk ordering an attack on the Klingons. The driver, Erik's brother, Neal, hunches over the steering wheel and warily guides the car down the rain-slicked road. There's no telling what monsters lurk ahead, concealed by the whipping curtains of water.

"Estimated wind speed: 50 to 60 knots," Rasmussen continues. "Very dark ahead. . . ." Directly in front of them, a brilliant lightning bolt slashes the sky in half. Trees flail in the high winds. At one second past 6:07 P.M., the Dryline Kid commands:

"Okay, now be real careful, Neal, slow down. This will be where we drive right into the tornado, so be very, very careful."

A moment passes, like the prelude to a punchline. Then— inaudibly on the videotape, but all too clear in Rasmussen's memory—someone in the back of the car says: "Look at that flock of birds."

Rasmussen turns to look. The "birds" are chunks of shattered houses hurtling through the air. He yells: _Debris! Debris! Tornado! Passing right in front of us!_ On the road ahead, the curtains of rain part, and the monster emerges. Everyone in the car gasps, freezes. A white, boiling tube of debris writhes across the road.

Moments later, the curtains of rain close again. The monster disappears into the deluge. Had they driven a few seconds longer, they might all be dead.

[Storm chasing] is one of the last frontiers for meteorology and mankind on earth.
—Storm chaser's comment on World Wide Web's
Storm Chaser Homepage (1995)

Tornado chasing is the logical culmination of a trend that began eight decades ago, when the Norwegian meteorologist Vilhelm Bjerknes urged his colleagues to get out of their offices and look at the sky. They spent too much time indoors, staring at barometers and thermometers and hygrometers, he complained: "We can learn much from [sea] pilots and lighthouse keepers that cannot be gotten at a desk." He believed weather systems possessed a hidden structure and choreography that could be divined by studying the arrangements and movements of clouds. His work led to the revolutionary concept of "fronts," that is, cold and warm fronts, which are now familiar on TV weather reports.

It was one thing to tell meteorologists that they should look at clouds. It was quite another thing to tell them to chase tornadoes! In the 1930s, the Estonian tornado scientist Johannes Peter Letzmann asked the U.S. Weather Bureau to start a tornado-watching program. A Weather Bureau official's reply could not have been more discouraging:

[Regarding] a tentative schedule for investigation of water-spouts and tornadoes proposed by Prof. Letzmann . . . we are convinced that it can never be fulfilled in this country. We have on the average nearly 150 tornadoes a year, scattered over a vast area, and only very occasionally at many year intervals is there opportunity for a meteorologist to actually observe an occurrence. They are so dreadful and destructive that usually the first thought of an untrained observer, unless sure that he is in a place of safety, is to seek cover for protection rather than make detailed observations as to the physical characteristics of the storm.

In all the years of record a tornado has never passed over a set of recording instruments, and in case one should, they would be demolished or carried away. People who have never witnessed an American tornado have no conception of its horror, and therefore it would be utterly impossible to make observations other than of a very general character.

Had that now-forgotten federal bureaucrat lived to 1995, he would have been astonished to see the VORTEX armada prowling the countryside. Its vans carried technologies unimaginable in the 1930s, including computer screens that displayed real-time satellite photos of a storm, radar units that scanned the storm's interior, and robot monitors that were sprinkled along the twister's expected path. And it was all funded by the federal government!

The National Severe Storms Laboratory (NSSL) is to tornado chasing what Florida's John F. Kennedy Space Center is to space exploration. It is located in Norman, Oklahoma, where scientific tornado chasing has been centered since the early 1970s. Norman is an urban speck surrounded by cattle country. In the Norman *Transcript*, lost-pet stories are front-page news; in motels, the clerks wear pink lipstick and beam like former cheerleaders. The wind sings through rusting oil rigs, mothballed until the next energy crisis. Family farms are disappearing. Downtown, many storefronts are boarded up. Ambitious young men and women—grandchildren of cowboys and wildcatters—graduate from the University of Oklahoma over on Boyd Street. Then they catch buses for Oklahoma City or New York or the West Coast, hoping to make it big like the state's other sons: Woody Guthrie, Ralph Ellison, Will Rogers, Tony Randall, Garth Brooks. Some environmentalists speculate about turning much of the state (and Great Plains) back over to nature: Declare the region a national park, turn loose the buffalo, and let prairie grass grow wild in ghost towns. The only remaining sound would be the Oklahoma wind, which never seems to stop blowing and, from time to time, explodes in unimaginable violence.

Two odd things happened in Norman in April 1995. At 9:02 A.M. on April 19, at the National Weather Service (NWS) office off Halley Road, the Doppler weather radar detected an unusual event in Oklahoma City, about 20 miles to the north. On the radar screen, a small blob swelled outward, like a balloon filling with air. Then the blob faded. Later, weather forecasters realized to their horror that the blob was an exploding building—the Alfred P. Murrah Federal Building in Oklahoma City, where, on that sunny morning, a terrorist bomb killed more than 160 men, women, and children. The authorities blamed the bombing on extreme-right-wing fanatics and drifters—the kind of people who come out of the woodwork when times are hard.

The other odd event also occurred off Halley Road, two days earlier, on April 17. A peculiar-looking armada of cars gathered in the parking lot by NSSL, a two-story, red-brick structure. Scientists milled around on the grass, conferring. The tops of twelve cars bristled with gadgets—wind vanes, temperature and air pressure sensors, and spinning anemometers. Two white vans gleamed with computers and radio gear. In the back of one van was a Doppler radar unit—white, dish-shaped, and as tall as a man. But this radar would look for tornadoes, not terrorists.

Eventually the scientists waved good-bye to each other (for the last time, if events that day went badly). They boarded their vehicles, started the engines, and—as neatly as a funeral procession—crept down Halley Road, turned onto Robinson Street, and accelerated toward the interstate highway. A year after its first campaign had ended in disappointment, VORTEX was on the road again.

In the distance, warm air rose (or "convected"). As it rose, the air lifted tons and tons of microscopic water vapor molecules. Air pressure drops with altitude; so as the air ascended, it expanded (as a sealed bag of peanuts expands when you drive up into the mountains). As the air expanded, it cooled; and as it cooled, its water vapor condensed into water droplets. These droplets reflected sunlight and became visible as puffy cumulus clouds.

On April 17, Rasmussen rode in one of the vans. He was the field coordinator or "FC," the guy who controls all the vehicles' movements by radio. The van was a high-tech spy's dream: "I'm in a van with a computer display that shows every trail, and dirt road, and paved road, and church cemetery, everything." On every chase, he would stay in constant touch with the other vans and with forecasters at NSSL and NWS. The forecasters monitored satellite and Doppler radar images of the storm and updated him on its movements. That way he always knew his team's position relative to the storm. A storm doesn't move in a straight line across the country. Like a giant amoeba, it unpredictably changes shape, surging here and there and sometimes changing direction. Once the VORTEX fleet neared the storm, Rasmussen would tell different teams where to position themselves around the thunderhead. The storm was constantly moving, so they'd pursue it like mosquitoes orbiting a fleeing rhinoceros. He'd regularly check back with them to ask their "azimuth"—their position relative to the tornado—so that he could plot the twister's path.

The VORTEX scientists' goal was to get as close as possible to tornadoes without dying. A close look was essential to resolve countless puzzles: How do tornadoes form? How fast are their winds? What is their internal structure? Why do only a small fraction of "supercells" (long-lived thunderstorms) give birth to tornadoes? Why do some tornadoes rage for hours while others dissipate in seconds? How can weather forecasters improve tornado forecasts? In 1994, they had listed seventeen tornado "hypotheses" they hoped to prove or disprove. By 1995 the list had grown to twenty-two. A neutral observer might have gotten the impression that the scientists weren't learning anything—that they were just getting more and more confused.

Confused or not, everyone had to keep his cool while chasing twisters. Cool heads avoided not only death but embarrassment. For in this age of commercially available police scanners and other off-the-shelf radio gizmos, someone might be listening in: "Control your ecstasy about seeing damaging weather (others are listening to our frequency and will see the tragic side of weather events), and do not use profanity. Talk professionally. Lighthearted giddiness may be good for relieving tension, but does not make a good impression on those listening in." These guidelines came from a widely distributed VORTEX memo.

Rasmussen was determined that no one would die on his watch. He insisted on knowing every vehicle's position at every moment. Otherwise, who knew what horrors would result? A VORTEX driver might chase a tornado down a one-way road, then have no escape route if the twister veered toward him. Tornadoes can turn cars into unrecognizable balls of steel; they'd have to pry out the driver's corpse with a crowbar. In short, VORTEX was a great deal of responsibility for Rasmussen, and he occasionally lost his cool. In his words, he would "go ballistic" from time to time. "This happens three or four times a day out there, [when] people aren't doing their jobs quite right and I get incredibly frustrated and start punching my fist into the roof and pounding on the table and swearing and cursing." He waves his gangly arms around to imitate a Rasmussen fit. He laughs about it now.

"In my whole life, I've never been high-strung or angry or violent. I'm just kind of a laid-back person. But when these things would happen, and I had 70 people out there, and all this time and money invested, and little stupid things would be screwing it up, I would just . . ." His voice trails off; he grimaces at the memories.

Staffers on different vehicles had different tasks. Some launched weather balloons, to measure the wind speed and direction at different heights. (Different wind velocities at different heights create "wind shear," which starts columns of air spinning and may spawn tornadoes.) Others recorded the twister (if any appeared) with 16 mm movie and video cameras. Their imagery would later be used to estimate the tornado's wind speed and dynamics.

Scientists also scanned storms with a special kind of radar called "Doppler." You've probably heard of Doppler radar in a less exalted context: Police use it to catch speeders. In meteorology, Doppler radar reveals whether atmospheric particles (such as rain) are moving toward or away from the radar. Scientists use Doppler radar to measure the speed and direction of airflow within storms (based on how air blows particles). They've also used portable Doppler units to measure the highest winds in tornadoes (287 miles per hour is the current record). During VORTEX, two aircraft flew around storms, Doppler-scanning their interior.*

Meanwhile, on the ground, Josh Wurman and Jerry Straka drove around in a van with a portable Doppler radar unit in back. The radar emits a radar beam with a three-centimeter wavelength, which allows it to see through heavy precipitation. As a tornado approached, they would drive off the road, open the back of the van, aim the radar, and scan the funnel and thundercloud. Their radar images show the tornadic funnel rising from the base of the thundercloud almost all the way to its top—a complete "body scan," as a doctor might call it.

Doppler radar works on this principle: A radar reflection from a moving object shifts in frequency as the object moves toward or away from the radar. If the object approaches the radar, the reflected radar waves compress together, so the radar receiver detects a higher-frequency signal. If the object recedes, the waves spread out, so the radar detects a lower frequency. The nineteenth-century Austrian physicist Christian Doppler demonstrated frequency shift in sound by placing musicians on a train and having the train move back and forth at high speed. As the train approached, the instruments' acoustic waves rose in pitch. As the train retreated, the pitch lowered. Likewise, when you stand next to the train tracks and a train approaches, its whistle rises in frequency or pitch. As it passes, the pitch falls to a moan.

Like Rasmussen, Straka looks younger than he is. He is physically compact, a thirty-three-year-old man with a youthful flippancy and a mischievous smile. Chasing "is mostly boring," Straka says with a smirk. "It's like high school sex—a lot of frustration and every once in a while you get lucky." Straka has been a meteorology professor at the University of Oklahoma since 1990. He claims to see tornadoes on one out of three chases, a remarkably high rate. (The much older and more experienced Professor Howard Bluestein claims only one in seven.) "You go out in the morning, see a tornado, go eat dinner at eleven o'clock at night, get home at six in morning. It sucks."

Still, Straka says, "I was *born* to do this. I've been interested in weather ever since I was young." Straka grew up in Wisconsin, near Milwaukee. His father drove him around the countryside so he could watch storms in action, and "I made my own little forecast every day." He recalls his first tornado: "There was a tornado watch. I went up on my roof, and a tornado passed almost exactly a mile from my house. It was hailing and raining on me."

He won his meteorology doctorate at the University of Wisconsin at Madison, where he also participated in stormchase teams. He met Kathy Kanak, another meteorology student; they later married. The couple chases together to this day.

Some VORTEX scientists pursued unusual projects. One, dubbed "In Search of Green Thunderstorms," would investigate the old legend that hail-bearing storms tend to be greenish. VORTEX researcher Frank Gallagher, working under William H. Beasley of the University of Oklahoma and Craig F. Bohren of Penn State, studied thunderstorm clouds with a spectrophotometer, a device that measures the exact frequencies (or colors) of light. Conceivably, hail-filled thunderstorms may be greenish because of the way in which descending hail refracts light.

One of VORTEX's most dangerous tasks involved "turtles." These are low, flat gadgets (hence the name) designed so that they won't flip over in high winds. They contain weather instruments. If a tornado loomed, brave staffers would race ahead of the twister and drop turtles along its likely path. With any luck, the tornado would pass directly over a turtle. If so, its instruments would measure the air-pressure drop within the funnel.

Until VORTEX, no one had ever obtained a convincing measurement of the air pressure within a funnel. The funnel is a rapidly rising, spinning column of air, and as it spins it centrifuges air outward, leaving a slight vacuum behind; thus the pressure drops. The lower the pressure drops in the funnel, the faster it sucks in new air through the bottom (where the centrifuging of air is reduced by friction with the ground) and the faster the funnel rotates. (It's like water going down a drain, except upward). In 1995, a tornado passed directly over a turtle for the first time.

Initially a tornado is invisible. You may see a spinning cloud of dust on the ground, but no funnel. The funnel appears only if its air pressure drops so low that inrushing air expands and cools enough to condense its water vapor into mist. The mist sheathes the cloud in a neat white column, peppered with whirling black rubble.

On April 17, storms brewed in north Texas. Thunderheads surged over the Red River and into Oklahoma. Inside his van, Rasmussen looked at his computer and studied weather maps and space photos, which he had downloaded by radio from NSSL and NWS. He concluded tornadoes were most likely to form that day near Temple, Oklahoma, just north of the Texas border. So the Dryline Kid ordered the VORTEX fleet to head toward Temple.

Along the road, on this and other VORTEX missions, Rasmussen kept noticing the same sight, a sight that has become painfully apparent over the years. Up and down the highway, everywhere he looked, were storm chasers—hundreds of them. He had mixed feelings on seeing them.

On the one hand, he was once one of them. Rasmussen spent the first twenty years of his life in Hutchinson, a town along the Arkansas River in central Kansas, where "you can't ignore the weather," he said. "I mean, it's just _dramatic,_ one day after another! Dramatically hot one day, dramatically cold the next. Flooding. I remember hail as big as softballs coming through the roof.

"I remember my neighbor's ham radio antenna getting hit by lightning. We were shocked—emotionally, you know—by this huge flash and bang. I think I jumped out of my chair. My neighbor claimed she saw 'balls' of lightning coming through her window.

"An incipient tornado came right over my parents' house before I was born. . . . A piece of debris came down through the roof and through the ceiling. Years afterward, I had that patch in the ceiling to look up at.

"I became a weather buff when I was little. I think it was in the second grade when the teacher had us put the forecast and the high- and low-pressure cells on a chart up on the wall. So I started watching the weather. . . . By junior high, I was _really_ interested in storms. I would climb a ladder up to the roof to watch them every time they started coming in. By high school, when I had my driver's license, I think one of the first things I did was to drive out and look at storms." He witnessed his first tornado in 1978. "I'd go out to the High Plains with the winds screaming and a huge CB [cumulonimbus or thundercloud] sitting there rotating, hailstorms coming down and popping you on the back of the head. . . . That was really fun."

No longer. "That part of the enjoyment has pretty much gone away. Either I've just gotten older, or there are so many geeks and nerds out there buzzing around and looking at storms that . . ." His voice trails off as he stares out an NSSL window at a clear blue Oklahoma sky.

"I think in the early days of chasing, there was this hunt-like aspect: You'd make your forecast, you'd go out to the blue skies just expecting to see something, and lo and behold, these rock-hard clouds would start building up and growing explosively. And slowly but surely, the storm would start to rotate, and a wall cloud would form, and then your tornado forms

. . . and you're sitting out there in middle of nowhere—you don't even know if there's any other humanity within miles—watching a tornado as the wind whistles through the wires and the birds are singing. That was really something. You'd gone on the hunt, and you'd got your prey, and you had captured it, and you were the *only* one to do it. That was really cool."

"But now," he says, sighing with a trace of resentment, "you drive out to a storm, and there's this *parking lot* out on the highway, with all these people with antennas sticking every which way off their cars and satellite dishes, and it's just . . . *aaahh!* Forget it!" He waves his hand in disgust.

"If I had to chase for pleasure nowadays, I would chase in Canada or Australia or someplace where it hasn't been done—to get away." Those countries have far fewer tornadoes than the United States, but "that's okay," he adds with a hopeful grin. "I'll still find 'em."

Some scientists criticize the recklessness of a few amateurs—that is, those who drive at 80 mph through a school zone, or cheer "Go baby go!" as a twister shreds a trailer park. "I try to distance myself from the amateurs," says NSSL's Robert Davies-Jones, one of the few chasers with white hair. In his soft English accent, he complains, "There's a fringe element that's just plain irresponsible. Most of them are responsible people, but there's an odd one or two who go around roadblocks, drive across farmers' fields, do crazy things."

By common agreement, the dumbest thing that a chaser can do is to "core punch." Core punching is the equivalent of strolling into a bear's cave with a bag of marshmallows. A chaser core punches by driving through severe rain and hail to get to the other, rain-free side of the storm, where the tornado usually forms. The danger is that you'll be so blinded by rain and hail that you won't see the twister until it lands on top of you.

Yet core punching appears to be an unofficial rite of passage for storm chasers, both scientists and amateurs. Everyone denounces it, yet everyone seems to have done it at one time or another. For every experienced chaser has faced at least one moment when he realizes: I'm going to lose this fast-moving storm if I don't take a shortcut. Unfortunately, the shortcut may lead through the storm core. But what the hell; the chaser gulps and grips the wheel and drives into the storm. Soon he's motoring through a back alley of hell: Rain drowns the windshield, baseball-sized hail crashes on the roof and cracks the glass, trees squirm and crackle and disintegrate. His eyes dart all around; frantic, he expects the twister to emerge from the torrent at any second, looming like a mountain and growling like an avalanche. But, chances are, he'll never see it. So he drives out of the storm, shaken and slightly disappointed—but *what a rush*! Afterward the sky clears, and a flood of cold, bracing air rushes in. He pulls off the road to relish the sight of the departing storm—a regal thunderhead with a wispy crown of ice crystals. It fades toward the east and glows amber in the sunset.

"Anyone who says they go chasing purely for science is lying," explains Straka, another ex-amateur-turned-scientist. "During a chase, when you see the expression on their face, you *know* what's driving them. . . . A lot of people I know dream about tornadoes all the time."

Now in his early fifties, Davies-Jones, one of the world's premier chasers, is expected to set an example for his youthful colleagues. He seems unusually dour for a chaser. Over a quarter of a century, he has witnessed enough twister mayhem to sober the craziest thrill-seeker. As he puts it, you wake up one spring morning, and "a lot of things go through your mind. . . . You get prepared to go out, and you may not be *feeling* like going out that day. And you get out [to the countryside], and you get anxious: Are there going to be storms? Are you going to be in the right spot? You start second-guessing what the field coordinator is doing. Anxiety mounts.

"Then when you actually *get* on [the track of] a storm, it's . . . it's . . ." His brow furls, he waves his hands; how can the English language express what he is trying to say? He looks like a man who has lived too long to talk about chasing as if it were merely fun. He knows what it really is: a perilous gamble that forces a man to acknowledge his mortality. "You have to constantly be *planning* what you're going to do. . . . One bad road decision takes you out of the chase."

Take the wrong turn, say, and you might drive into a hailstorm. Midwestern hail isn't like the wimpy, pea-sized hail of the East and West coasts. Midwestern hail is jagged and as big as golfballs—baseballs! It slaughters farm animals and pummels a car all the way to the junkyard. In 1995, hail would cost Davies-Jones "two windshields—front and back. The first big hailstone made this big 'spider' crack. Then some of the other ones—four or five big hailstones in all—made these big indentations in the windows. There were big bulges and dips in the glass. We kept on driving anyway. . . . Little slivers of glass were everywhere. My driver got slivers in his face, I got them in my laptop computer while I was typing on it. . . . I put my clipboard against the windshield for some protection."

On April 17, as the VORTEX caravan rolled west, the sun continued to shine, heating the terrain and inciting more convection. A few cumuli swelled into towering "cumulus congestus" clouds, which resembled aerial ski slopes. And some boiled into thunderclouds, whose crystalline tops brushed the sky's "ceiling"—the tropopause, gateway to the stratosphere. Their broad bottoms were as black as night. And if the scientists were lucky, those bottoms would soon bristle with twisters.

Davies-Jones and his team rode in "Probe One" of the VORTEX fleet. As Probe One neared its destination—Temple, Oklahoma—Davies-Jones's heart soared at the sight of a "glorious supercell" on the horizon.

Then his car broke down. Damn. Of all times. The driver pulled to the side of the road. Davies-Jones used his cellular phone to summon a tow truck. Then they waited . . . and waited . . . and waited. Meanwhile, Rasmussen and the other scientists drove on to Temple. As minutes passed, Davies-Jones listened on his radio to their crackling voices as they approached the Oklahoma-Texas border. They began to sound excited, confused, frightened. Something big was cooking near Temple.

Something big, indeed. "At one point," Rasmussen recalls, "we were just southeast of the reported radar location of the mesocyclone, by about five miles." (Again, a mesocyclone is a rotating, vertical column of air, typically a few miles wide, within a thunderstorm. Tornadoes may descend from a mesocyclone.) He had to decide which route to take next. The decision wasn't easy. Normally chasers head for the southwest quadrant of a storm; that's where tornadoes are likeliest to sprout from the rain-free base of the supercell. Then they would normally watch for the descent of the "wall cloud"—a dark, visibly rotating cloud that drops from the base of the thunderhead. The wall cloud is the lower, visible portion of the mesocyclone. When the wall cloud appears, it's showtime: A tornado may drop at any moment.

But this storm was an enigma: "It didn't have a wall cloud. It didn't have *any* of the cloud features you normally associate with a mesocyclone." So where should he tell the fleet to go? He looked at the sky, then studied his weather and road maps, then looked at the sky some more. . . .

"We were looking northwest, and we saw this big, low-hanging cloud draped down to the south and southwest. It looked like a gust front—cold air outflow from the storm. Looking at the map, I could tell that one road option was to go north in front of this gust front and then go east and stay just right ahead of the storm." That might be risky: A mean gust front can knock a car off the road.

"The other road option went southeast and east, and I knew that if we took that, we would have lost the storm and never caught it again. And it was really the only storm out there; it was moving fairly quickly, and we had to keep jumping to keep up with it.

"So I made the decision to go north and east, thinking that we'd probably end up slicing through this gust front and then coming out ahead of it again. We'd encounter some gusty winds and a little battering, but nothing real *hazardous*.

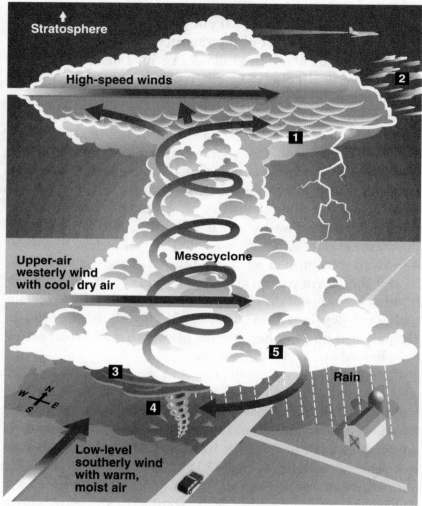

Joe Shoulak

1 Rising warm air condenses and forms bumpy mammatocumulus clouds under the thundercloud's "anvil."

2 High-speed winds freeze water vapor, creating wispy, crystalline cirrus clouds atop the storm's "anvil."

3 Wall cloud.

4 Tornado.

5 Rain causes cold downdraft that either (a) snuffs out the mesocyclone tornado or (b) generates its own low-level wind shear, forming a new tornado.

"So we started heading north in a pretty tight group, maybe 100 yards between each car. All of a sudden people starting yelling on the radio that there's a tornado. I'm looking all around and see no tornado. I'm thinking, 'What are they talking about?' " He yelled into his radio microphone: "*Somebody* tell me what they're talking about!"

"Then the people in *my* car started yelling, 'There's a tornado, there's a tornado!' " He looked around again, still saw nothing, and barked: "*Will you people tell me what you're talking about!?*"

"And all of a sudden, about 100 meters west of us in the field, I see this dust whirl start to form—a real tight little one. I thought: '*Oh shit*—it's *right here.*' They should have been yelling: 'There's a funnel *overhead.*'

"I knew the tornado would be moving roughly east. So I got on the radio, and I yelled, '*Tornado just west of the road. The vehicles in front of the FC [Rasmussen] should go north as fast as you can. Vehicles behind FC, stop now and let the tornado cross road in front of you.*' That would make a 'hole' in the caravan for the tornado to go through.

"I jumped up in my seat and watched out the back window as the tornado crossed the road. I was just freaked because the vehicle behind me was a group that liked to stay with me. They felt insecure if they weren't right on my bumper. So they didn't follow my instructions; they decided to go ahead and follow us down the road. I saw this tornado coming at their vehicle, and I thought: '*Oh shit, it's going to take them out.*' "

The twister ripped across the road. An explosion of dust and debris blocked Rasmussen's view of the car behind him.

In another car, Straka watched the tornado approach the road. The funnel was small but furious. He was sure it would hit some cars, "and that would be the end of VORTEX." By radio to Rasmussen, Straka read off the distances between the twister and the rear car: "100 feet . . . 50 feet . . ." The funnel passed within a stone's throw of the rear car's bumper. Straka shouted into his mike: "No cars in the air!"

That type of close call frayed Rasmussen's nerves. And the

chase season was just beginning; what would the rest of the season bring?

Miles away, Davies-Jones and his team fumed by their dead sedan. They had spent hours waiting for the tow truck. Tumbleweeds jiggled across the prairie. Cars whizzed by indifferently.

Finally the tow truck arrived. After their vehicle was repaired, Davies-Jones and crew glumly drove toward a motel, figuring they had missed all the fun.

But not quite. "Near dark," he recalls, "this other line of thunderstorms developed to our west, and Bill, my driver, said: 'Boy, that looks like a wall cloud, if I didn't know any better.' It turned out it *was* a wall cloud, headed right toward us. It passed just to our north.

"This wind starts blowing about 70 mph, and each gust seemed to be higher than the one before it." Visibility fell to zero. "I really thought about getting out of the car and hitting the ditch because I thought our car was going to be blown across the road."

They never saw any tornado, and the turmoil lasted just a minute. "A scary minute." The brief blast lifted their spirits, and they continued to a motel, tired but happy.

VORTEX's research is important, and not simply because it uncovers new facts to fill overpriced textbooks.

Over the next century or so, tornadoes may become the commonest force of urban devastation in the United States. The reason isn't because tornadoes are "on the march," however. *We're* the ones on the march.

Until now, America's high tornado rate, while tragic, has been societally tolerable for a single reason—a reason clear to anyone who has flown over the United States at night. Look out the airplane window and what do you see? An immense blackness. Occasionally a glowing patch floats by—a village, town, or city; then the blackness returns, sometimes for minutes at a time.

The reality is that the United States is remarkably underpopulated, as nations go. And luckily so; if we were as densely populated as most other countries, our annual tornado death rate would be gruesomely higher. Consider that the population density of France is 275 people per square mile; China, 322; Germany, 588; the United Kingdom, 616; Japan, 857. But the United States is a mere 70 people per square mile. Therefore most U.S. tornadoes live their short lives without hurting anyone or damaging anything—or, for that matter, without even being seen. Tornado climatologist Thomas P. Grazulis estimates that only one out of three American tornadoes is observed and recorded.

This cannot last forever, of course, for our population is growing. The U.S. population was 76 million in 1900, 151 million in 1950, and 249 million in 1990. It is likely to surpass 300 million by the middle of the next century. The U.S. population density (people per square mile) has doubled since the 1930s—the length of a single human lifetime—and will continue to rise even if family sizes remain small (one or two kids per couple). Those extra souls will have to live somewhere, and if current trends are any sign, most of them will choose *not* to dwell in densely crowded, polluted, crime-ridden, traffic-jammed cities. No, they'll want "leg room." So they'll leave town for the suburbs—or beyond. That is why so many towns and cities are already spreading out like syrup on pancakes.

As cities expand, they become bigger and bigger targets for passing twisters. Therefore the annual death rate from tornadoes, which has plummeted since the 1970s (at least partly thanks to better forecasts and preparedness), may one day rise again.

"A tornado striking the center of a large city would certainly be a major disaster," a National Academy of Sciences study said in 1987. Allen Pearson, former director of the National Severe Storms Forecast Center (NSSFC), once cited his "own private hell"—the fear that a tornado will hit a huge gathering of people, such as a sports stadium, rock concert, fairground, or the Indianapolis 500.

The tornado rate is also likely to rise because of *where* people are moving. An astonishing number of Americans are (unwittingly, one presumes) moving to tornado-infested regions, notably the Southeastern states and Florida. U.S. Department of Commerce figures show that the 25 U.S. counties with the highest population growth from 1980 to 1992 included 16 in states that are especially vulnerable to tornadoes for one reason or another. Those 16 counties include five in Florida, five in Texas, five in Georgia, and one in Arkansas. These shifts are troubling because tornadoes are major troublemakers in those states. Florida has more reported tornadoes per 10,000 square miles than *any* state in the union. Arkansas has the most "killer" tornadoes—i.e., twisters that kill at least one person apiece—per 10,000 square miles of any state. Texas has both the most tornadoes and tornado-related deaths. And Georgia ranks seventh in total tornado deaths from 1880 to 1991.

In contrast, the bulk of population is generally *leaving* states that are relatively safe from tornadoes. Of the 25 counties with the biggest population declines, only 6 are in states where twisters should be a significant worry (Illinois, Missouri and Louisiana).

Some Americans are abandoning community life altogether. They are moving to rural Erewhons where (they hope) they will enjoy a quieter, crime-free life, linked to the world via computer networks and satellite dishes. Unbeknownst to them, their nearest neighbors may be various atmospheric vortices, from dust devils to full-fledged twisters. Over the next century their electronic outposts will expand into villages, then towns and cities—each one a potential target for a tornado. As America grows, so does the number of targets.

This spreading population partly explains why the reported tornado rate has skyrocketed over the last 40 years. A century ago, official records listed fewer than 100 tornadoes a year. In 1950 alone, the number was about 200. From 1953 to

1991, the yearly average was 768. In the 1990s, the annual rate has surged above 1,000, and it approached 1,300 in 1992. As the population spreads outward, tornadoes are less and less likely to pass unnoticed.

This intimate link between tornado sightings and demography also makes it difficult to say where "Tornado Alley" really is. Many or most people assume the "Alley" is—as the names implies—a narrow, well-defined strip through Kansas and a few other unlucky farm states. In reality, when it comes to tornadoes, much of the United States is ground zero.

We can't even clearly identify one state as the "tornado capitol." The tornado climatologist Thomas P. Grazulis points out that more than a dozen states are candidates for "capitol" in one respect or another. It all depends on how you define "capitol": Is it the state with the most tornadoes? or the most tornadoes per unit area? or the most deaths per tornado? or some other criterion?

For example, you'll probably never guess which state has the highest average number of tornado deaths per 10,000 square miles. Massachusetts! Do you know which state has the highest concentration of tornadoes that result in injury? Delaware! "Tornadoes are nature's most violent storms. . . . Every state is at some risk from this hazard," according to the Federal Emergency Management Agency (FEMA).

To use the crudest form of comparison, let's look at the total number of tornadoes reported by state from 1953 to 1991. By that criterion, the "capitol" is Texas, with 4,949 tornadoes (an average of more than 100 per year). The rest of the "top 10" list are Oklahoma (2,042), Florida (1,762), Kansas (1,736), Nebraska (1,449), Iowa (1,229), Missouri (1,044), Illinois (1,042), South Dakota (970), and Louisiana (949). Texas also had the highest number of tornado-caused deaths (448).

But this "top 10" list is misleading because states vary in area. Texas has many twisters partly because it's a gigantic state. So it's more meaningful to rank states by the number of reported tornadoes *per unit area*. In that regard, the "capitol" is Florida, with an annual average of 7.68 reported tornadoes per 10,000 square miles (that is, an area equal to a square of 100 by 100 miles). Next in line are Oklahoma (7.44), Indiana (5.51), Iowa (5.50), Kansas (5.35), Delaware (5.00), Louisiana (4.95), Mississippi (4.82), Nebraska (4.79), and Texas (4.71).

Many Americans mistakenly think their state "never" has tornadoes. This writer lives in California, where most people would sooner expect a Martian invasion than a twister. Earthquakes, mudslides, brushfires—yes! But tornadoes? Never! Indeed, California ranks low (32nd) in the total number of recorded tornadoes from 1880 to 1991. Yet the grim truth is that more than 80—that's right, 80—tornadoes have hit the Los Angeles basin alone since 1962. They tend to be weak, but even a weak twister may cause costly damage.

Not all tornado chasers are in the Midwest. A small number of amateur chasers pursue tornadoes in Eastern and far Western states—from California to Mississippi to upstate New York to the outskirts of Washington, D.C. One is Steve Johnson of Fresno, California. He is a wholesale distributor of beauty products who got his degree in meteorology and chases tornadoes in California's Central Valley. The Central Valley is California's main cropland, a sort of Midwest in miniature, sandwiched between the Pacific coastal mountains and the Sierra Nevada range. By early 1995, Johnson had chased 16 Central Valley funnels and videotaped seven.

"We need more tornado watchers in California to correct the misimpression that there are no tornadoes in California!" Johnson says. "That's *wrong!* . . . In March 1991 I was *underneath* a tornado at Chowchilla [California]. It was a close call. A very heavy wind buffeted the car and almost knocked me to the ground." The funnel passed within 500 feet of him; it flicked a tractor 100 feet. "If I close my eyes I can still hear the sound. It was something I'll never forget—like a high-pitched whistling noise and a low rumble at the same time."

Why does he risk his life chasing funnels? "It's a *passion*," he declared, thumping his chest with his fist. "No matter what it costs, it's a true passion. Sometimes I put 300, 400 miles on my car in a [chase] day. . . . [Chasing] prevails over my business, it prevails over my personal life. . . .

"All my life I've thought of severe storms as God and Mother Nature's most awesome examples of sheer beauty. I want to know what makes Them tick—and the tornado is only the tip of the iceberg."

Shades of Ahab: The best chases have been compared to religious experiences—to encounters with the infinite. Greg Stumpf is thirty-two, tall and sturdy-looking, a native of Poughkeepsie, New York. He works in Norman, Oklahoma, at NSSL and the University of Oklahoma, where, among other things, he's trying to teach computers how to identify mesocyclones in radar images. Very adult stuff. Yet his diaries sparkle with a youthful, shameless thrill:

May 11, 1992, south-central Oklahoma

. . . a mesocyclone formed right on Route 32 . . . all of the sudden we began to see DEBRIS in the air right down the road! We saw pieces of metal scaffolding, and tree branches just sort of floating there, but going around in a cyclonic circle. We then saw an entire TREE with the root ball attached to it still about 200 feet in the air!!!

March 21, 1991, near Atoka, Oklahoma

To the west, the sky appeared to be growing darker. . . . Within tens of seconds, the tornado funnel condensed to the ground, quickly becoming your classic WEDGE shape! It was directly to our left, rotating a lot, and bearing upon us fast. It was also still raining and hailing quite hard. It was IN-CREDIBLE! The tornado was wider than it was tall. It was spinning quite hard. There were multiple vortices going around it. You could see condensation debris rolling up the north side and sucking in at ground level. And it was getting close! Dave Gold reached for the video camera and started filming out the back of Gene [Rhoden]'s car. The wedge was *right on our tail, and we knew it, so we started driving about 55 mph through the rain and hail to get away. I kept looking back at it, and looking forward on the road to make sure Gene was not going to kill us in a car accident. All of the sudden, an 80 mph gust blasted us from the northeast, the rain stopped temporarily, and dust and/or spray blew across the road, bucking the car for about 10 seconds. On video, you can hear me shout "Holy Shit," and you can hear the wind hit the car. I thought it was the end. . . . Within the next 30 seconds, the tornado crossed the road about ¼ to ½ mile behind us, and became lost in the precip[itation] and trees to our south and southeast.*

April 12, 1991, near Enid, Oklahoma

. . . a new wall cloud began to form rapidly just to our NNW. It was front-lit, so it appeared white on a blackish background. We drove north, and stopped less than ONE MILE from the edge of the wall cloud. Let me tell you, this was the most INCREDIBLE rotation I have ever seen!!!!! It totally blows away ANYTHING I have ever seen. We go SO CLOSE! The wall cloud was almost smooth on the outer edge, and looked like the edge of a wedge (pardon the rhyme) . . . [A funnel descended] and a small debris cloud was seen. The motion in the wall cloud and funnel were TREMENDOUS! I saw two [electric power line] transformers flash in the debris cloud. . . . It destroyed a machine shed and uprooted a railroad crossing signal.

Then they chased another tornado, which descended nearby.

I got out of the car, ran across the highway with my camera, and stood next to the field it was in. I COULD HEAR THE SOUND AND NOISE!!! It was like the sound of a distant waterfall. . . . Not really a roar, but a gushing sound. YESYES-YESYESYES!!!!! It was a CLASSIC DRILL PRESS! GAWD!

But "encounters with the infinite" were merely memories to Stumpf and other VORTEX scientists through most of the spring of 1995, when "good" tornadoes were few and far between.

The weather was only partly to blame. True, big tornadoes were rare, but hardly nonexistent. A terrific twister struck Ardmore, Oklahoma, on May 7. It killed three people and smashed a Uniroyal tire plant. Police officer Darren Culley saw the funnel approaching Plainview High School about 5:00 P.M.: "I knew it was severe by looking at the size of it," he told the *Daily Oklahoman*. "Then it landed right on top of me. . . . I saw the roof of the high school start to lift up. I heard whistling. The pressure in my car built up, so much my ears went to popping." His car briefly lurched from the ground. Tamyra Jones heard a radio report about the funnel and herded her two kids into the bathroom "and prayed . . . We could hear the whole house and furniture [hurtling] against the walls." Her roof ripped off. The bathroom was the only place in the house that survived intact. The tornado threw her van into the backyard, killing her dog, Marbles. At the Uniroyal plant, 500 workers rushed into shelters. The lights died and their ears popped. They could hear the plant "ripping all to pieces," just outside the shelter door.

Unfortunately, VORTEX missed the Ardmore storm. The VORTEX fleet was busy at another storm that day, and the VORTEX rule was: Stick to the storm you're with; don't go running off after another one just because it looks better. Otherwise, VORTEX might never achieve one of its goals: to study the evolution of tornadoes through their entire life cycle, from birth to death. Still, it hurt to miss a beauty like Ardmore. "Clearly," Rasmussen and Straka acknowledged in a memo, "the Ardmore storm would have made a much more interesting target."

By the start of June 1995, days before VORTEX's scheduled shutdown, "we didn't have a whole lot of very exciting data," Davies-Jones admits. It looked as if the project would end—its final day was June 15—with little to show for everyone's toil. By early June, the VORTEX scientists were wearing out. VORTEX scientist John T. Snow of the University of

Oklahoma grew worried about the researchers' "sleep deprivation [and] accumulated fatigue. They had gone on for, I guess, ten weeks by that time." He jokes: "I'm just glad our university risk managers weren't watching." Hail damaged several cars, and "there were several minor injuries. We did have one aircraft incident where an aircraft hit a downdraft fairly close to the ground and a lot of people went floating around inside the airframe. People hit their heads."

On the evening of June 1, Rasmussen studied the weather charts and concluded that storms would soon be brewing over Clovis, New Mexico. Clovis was 100 miles northwest of Lubbock, Texas, where he had attended Texas Tech in the early 1980s. "I looked at the forecast for the Lubbock winds, and they were forecast to be out of the southeast at 30 mph and unstable. I thought, 'Wow, when I lived there, that always meant big stuff.' So I was excited the night before.

"I came in next morning. Nothing had changed much. It still looked like there was going to be this strong, low-level southeast winds and big instability, and I thought, 'It looks like [we'll have] a really typical west Texas tornado outbreak.'

"Usually in the morning Jerry Straka comes in, and he's the devil's advocate: He tries to tell me all the reasons why something *isn't* going to happen. But that morning, I was so determined to get everyone out there [to west Texas] that he just left me alone. So we got out of here at 10 A.M.

"Sure enough, we got down to west Texas, almost to Clovis, and these *big, hard* clouds were going up in eastern New Mexico!" The clouds resembled mountains of whipped cream. The higher they rose, the darker their bases became. Their tops soon bumped against the tropopause. At those ethereal heights, cold, high-speed winds froze the clouds' water vapor into ice crystals and fanned them outward, forming classic thunderhead "anvils."

Time passed; the thunderheads evolved very slowly. Rasmussen grew antsy. Why weren't the thunderheads turning into thunderstorms? "It made me angry. I thought, 'We're going to get screwed again.'"

Eventually a few storms intensified, especially one near Clovis, "and we started following it northeastward. . . . The inflow wind got so strong and the dust got so thick that we couldn't see the cars in front of us. Some of the power lines were being torn down. I thought, 'No matter what we see today, it's going be something exciting, because you don't get that kind of inflow wind on an average storm.'

"Shortly thereafter, we started getting reports of very intense rotation on the Doppler radar. Jerry called in from the mobile Doppler and said he'd seen all kinds of 'folds' in the wind velocity. That means the velocity is stronger than what the radar resolves. When that happens, you know the storm is getting stronger and stronger. Also, Probe One and a few other Probes started reporting strong rotation and a bit of blowing dust.

"When that happens, I kick into 'robot' mode—calling people, finding out where they are and what they're seeing, triangulating it on the map, relaying the projected positions back out to the teams; that just goes on continually. Once that happens, there's really no time to get excited or enjoy the storm; I'm just pretty much all business."

The real action began just southwest of the small Texas panhandle town of Friona, about 30 miles northeast of Clovis. A violent twister formed southwest of the town, passed on the southern outskirts and through the eastern side of the community. It destroyed a large grain elevator and wrecked the local airport, including a large steel building. "The anchor bolts holding the columns of this building were ripped out of the concrete slab, with part of the slab going with the bolts," Rasmussen and Straka said. "The heavy beams were left in a twisted heap. One I-beam became a missile and was thrown about 100 meters." The tornado picked up a railroad boxcar weighing several tons and bounced it for 100 meters through a cemetery, smashing monuments and digging a two-foot-deep hole into an asphalt road. Then the Friona tornado disappeared into heavy rain. Throughout the storm, the VOR-TEX vehicles raced around the twister, scanning it with

Doppler radar, dropping turtles, and launching unmanned balloons with weather instruments and radio transmitters.

"There were at least six rotating supercells in a three-county area," Rasmussen recalls. "The sky was almost black as night. Blowing dust everywhere. I felt like the atmosphere had gone completely berserk."

Then he got a radio report that yet another storm was forming 20 miles to the east, near a flyspeck of a town called Dimmitt. He ordered his team to head there. He felt both happy and unhappy: happy because at Friona, they had finally cornered a big tornado; unhappy "because I knew we hadn't gotten real comprehensive data" there. Dimmitt offered a second chance.

"We get down to Dimmitt, turn east, and here's this big supercell bearing down on the town with a big rotating updraft." He observed a "clear slot," an opening in the storm caused by descending drier air. He got on the radio and announced that he expected to see "a significant tornado in the next few minutes." The Dryline Kid was right again: Soon a fat funnel cloud began to descend.

The tornado hit the south side of Dimmitt. It crossed State Road 86 and ripped off hundreds of square meters of road and spewed the black asphalt more than 600 feet into a field. Two truck-trailers were blown away.

Rasmussen was 6 miles east of Dimmitt. He had a beautiful view of the tornado. He tried to radio to his troops who were closer to the storm. He tried and tried. No one answered. They were watching the tornado.

"What happened at Dimmitt was that the tornado formed and"—Rasmussen's voice softens, his eyes get a faraway look—"everybody was just *mesmerized*. I mean, it was a *beautiful* sight. And for about 40 seconds, I couldn't get anybody to tell me what the azimuth of the tornado was from their vehicle. . . . No one would tell me anything."

That was too much for Rasmussen. After two years of repeated frustration, here they were, within the equivalent of a few city blocks of the mother of all **VORTEX** tornadoes, and he couldn't get anyone to pick up their blasted microphones. He gleefully imitates the tantrum he threw—yelling, fists smacking the roof—as this greatest of VORTEX twisters roared by.

As it turned out, Dimmitt was a historic tornado. It was the first major twister that VORTEX scientists documented from beginning to end. Radar probed its innards in unprecedented detail; movie film and videotape recorded its mad whirrings.

For years to come, the scientists will pore over the Dimmitt data and plumb its secrets. It will be used to test theories of twister formation and, perhaps, to find better ways to forecast them. VORTEX, which could have ended in failure, climaxed in triumph at Dimmitt, where even the coolest scientists momentarily forgot themselves and, jaws hanging and arms limp, gazed in awe at the whirling leviathan. "Beautiful," they whispered to themselves, over and over and over. And beautiful it was—as beautiful, in its own way, as the Grand Canyon or a solar eclipse. It's no wonder that the more mystically minded chasers have perceived, behind the serpentine immensity of a twister, the shadowy outlines of transcendental forces, perhaps even of God.

Enraptured by this aerial vision, anyone might forget what a tornado really is: an engine of annihilation that can wipe a town from the face of the Earth. Literally.

CHAPTER 2

A World Destroyed

*A*fter midnight on June 8, 1984, a twister demolished the sleeping village of Barneveld, Wisconsin. Virtually nothing was left standing. Some residents thought a nuclear bomb had fallen. The death toll of nine was small as major tornadoes go. Yet Barneveld was so tiny, its obliteration so total, and its people such evocative symbols of an American ideal— small town life, far from the pressures and alienation of the city—that its fate made global headlines.

The Barneveld disaster is scientifically important because it inspired a new area of tornado research, long-distance debris dispersal. The tornado blew debris for extraordinary distances, so extraordinary that it raised an unsettling question: What would happen if a tornado hit a chemical plant, radioactive waste site, or other environmentally "sensitive" facility and scattered its toxins for tens or hundreds of miles?

The tragedy also shattered one of the commonest American delusions about tornadoes: that a town can be "protected" from twisters by a local quirk of topography. Barnevelders regarded a local hillside as their "shield" against tornadoes—a paper shield, as it turned out. Across America today, residents of countless other communities suffer similar delusions of immunity from the atmosphere's cruelest storm. If history is any guide, there is only one cure for this delusion: a visit from a tornado. Its visit exacts not only a physical price, but a psychological one. After the winds subside, townspeople may rebuild their homes and businesses and schools. But to survivors, the sky may never look so blue and vast and hopeful again; they may perceive a dark lining in every silver cloud.

If you drive west from Madison, Wisconsin, on Highway 18-151, you pass through farm country. Tractors plow fields; cows meander; a white, spotted horse looks sleepily out of a stall. Just past the sign for "Lost River Cave," turn right on the "K" road, pass the old cemetery, and you're in Barneveld. It looks like a Walt Disney movie's concept of a small town: Cozy little homes, a church steeple here and there, a corner bank where the tellers know the customers' first names. Over it all looms a white water tower that proclaims "BARNEVELD." But the homes look eerily new; the gas station and grocery store seem too big and shiny for a remote country village of 800-odd souls. Travelers who stop for gas or coffee ask locals the same question: What is such a modern-looking community doing in the middle of nowhere? "They don't know there was a tornado," observes resident Anita Jabs.

Long ago, on spring and summer evenings, songbirds chirped in Barneveld's century-old oak trees as the shadows grew long. On Jenniton Avenue lived Mary Ann Myers, a petite, cheerful-looking widow with her white hair in a bun. She liked to sit outside in the evening and watch the western horizon. There, incoming thunderstorms flickered with lightning. Immense cold fronts surged out of Canada or the Rockies and passed over southern Wisconsin en route to the Great Lakes, where they occasionally spawned waterspouts before drenching Chicago and points east.

A few locals feared that one of those storms would drop a tornado on Barneveld. But most residents laughed. Don't worry, they said; Blue Mound will protect us. Blue Mound— named for the blue haze around it—is a 1,716-foot-high hill near Barneveld. From the top, one can see the State Capitol in downtown Madison, 25 miles to the east. To residents, Blue Mound was their shield against tornadoes—a physical barrier that would block any twisters. Their view made no scientific sense, but it comforted them.

Besides, it was hard to imagine disaster striking Barneveld, where *nothing* much ever happened. In 1899, President McKinley's train stopped at the local station and he emerged to give a five-minute speech, and in 1910 the Congregational church burned down; but that was about it. Barnevelders were descendants of lead miners and cattle and dairy farmers who, a century earlier, had fled European poverty for America. They were quiet, self-effacing folk from Wales, Ireland, England, Scotland, Germany, Switzerland, and especially Norway. Religious, too: Barneveld's three churches (Lutheran, Catholic, and Congregational) were packed every Sunday. With God and Blue Mound to protect them, what did they have to fear?

June 7, 1984, was "a hot sticky night, a real true summer night," Mary Ann recalls. To the west, in Iowa on the other side of the Mississippi River, storms were churning and tornadoes touched down. Mary Ann had to travel to a wedding the next day, so she stayed up late that night paying bills, balancing her checkbook, and writing notes to her family. Nowadays, she wonders if she did all those things because she had a "premonition" of coming events. Anyway, she penned the last note and went to bed. Also sleeping in her two-story duplex were her two daughters, Anita and Jill, Anita's son, Joey, and the elderly mother of Mary Ann's late husband. Usually Anita put four-month-old Joey in his crib at bedtime. But that night, for some reason, Anita took Joey to bed with her.

The lights winked out across Barneveld. Anita Jabs, who ran the local store with her husband, Ron, had heard about the Iowa tornadoes and watched TV until 11:00 P.M., waiting for a tornado warning. None came. So she hit the sack. Elsewhere in town, two young parents, Charles and Susan Aschliman, tucked their two-year-old boy, Matthew, into bed. After a day spent planting flowers in their garden, Roger and Jeanne Jabs turned out the lights; the only sounds in the house were the panting of their dog, Belle, and the purring of the cat, Kalamazoo. Over in Barneveld's handsome new Thoni subdivision, James Slewitzke prepared for bed. A resident of northern Wisconsin, he was visiting Barneveld to help his sister Elaine paint her house.

Nearby, Ruth and Dick Ehlert stayed up late, preparing for a family car trip to California. Dick was Barneveld's only police officer; Ruth was a legal secretary. They had married the previous October, she for the second time. They would make the California trip with her teenage sons, Keith and Kyle. Keith still lived with his father following the breakup. Ruth "hoped desperately" that the trip would bring her and Keith closer together. With the long journey ahead, Dick suggested they hit the road that night. But Ruth was weary. No, she said; let me get a few hours' sleep. By 12:50 A.M. on June 8, 1984, she was in bed.

A few miles away in the town of Black Earth, officer Bob Schuyler's police car crept through the empty streets. Black Earth wasn't exactly a hotbed of crime. But even small-town cops keep their eyes peeled for high school vandals, especially with summer vacation on the way.

Sometime after midnight, 30 miles southwest of Barneveld, a tornado touched down near the village of Belmont. It was a small twister—an "F2" tornado, about 120 feet wide. ("F2" represents a "significant" tornado on the Fujita Scale, which investigators use to estimate the intensity of tornadoes based mainly on building damage.) The Belmont twister was the first of several tornadoes that would whir across southern Wisconsin in the next hour or so.

At 12:50 A.M.—that's when everyone's electric clocks stopped—a "horrendous wind" awoke Anita Jabs. She turned over in bed and punched her husband, Ron. "I'm going to the basement," she snapped. "You get Jason and I'll get the girls."

Nearby, the sleeping Aschlimans were startled by two-year-old Matthew's wails. Susan grabbed Matthew, and the family fled to the basement. As they reached the basement stairs, they heard a harsh sound overhead, the sound of shredding wood.

At Jeanne Jabs's home, she left bed to close a window. Then the wall disintegrated. She seized her cat, Kalamazoo, and hid behind a chair.

Another Barnevelder, Doug Salisbury, frantically raced through his home and fell on top of his daughter, Megan. "Daddy, Daddy, get off of me—I can't breathe!" she screamed.

A spectacular blast of lightning awoke Mary Ann Myers. She ran to her bedroom door to tell her family to hurry to the basement. She opened the door—and stopped dead in her tracks. The other side of the house was gone.

"I was looking outside at the ginkgo tree, blowing in the wind. . . . I died a little then," she recalls. "I couldn't breathe—I wonder if it was the shock or the air pressure. I remember screaming. I had two daughters and a grandson on the other side of the house, and a mother-in-law down below, and I told myself: 'They aren't there anymore.' I screamed and prayed to God, all at the same time."

"An enormous roar" shook the Ehlert home. With a cop's instincts, Dick reached for his portable radio. Then "all hell broke loose," Ruth later recalled. "The windows blew in, glass shattered everywhere, the roof flew off, some of the walls collapsed." Rain and dirt gushed down. Something heavy dropped on Ruth. She flailed on the floor, yelling to Dick and demanding to know if their sons were okay. "Make them talk to me!" she screamed. The boys answered; they were alive. Dick Ehlert started to radio for help. Then he discovered the radio was covered with a wet, dark substance: mud. Furious, he slammed the radio against the wall. "Don't!" Ruth implored. "That's all we've got!" He picked up the radio again and broadcast a call for help—help from anyone, anywhere.

Across Barneveld, men, women, and children crouched in their shattered homes and listened to the torrent fade toward the northeast, toward Black Earth. Susan Aschliman stared at her empty arms, which a moment earlier had held little Matthew. He was gone. Something had knocked him from her grasp. He was somewhere outside, out in the darkness and screeching wind.

Jeanne Jabs awoke in a black, wet field. She wondered: *How did I get here?* Then she realized she was across the street from her house. Roger was still in the house, groggily peeling insulation from his eyes and mouth and thinking, "Okay, when am I going to wake up?" He heard their dog, Belle: She whined and walked atop debris that had crashed from the ceiling.

Numerous other Barnevelders were blown from their homes. One second they were fast asleep, warm under the covers while rain drilled the roof and winds shook the windowpanes. The next second they awoke in their pajamas, sprawled in a muddy field hundreds of feet from their beds. They felt wet. Were they soaked by the rain? Or bleeding to death? They couldn't tell; the gale had knocked out every light in town.

Some suspected it was the end of the world. "Could this have been a nuclear war?" Roger Jabs wondered. To Cindy Schaller of Jenniton Avenue, the wind was "like a huge force pulling at you, just like it was going to pull you apart. . . . For a very fleeting instant I was afraid it was a nuclear bomb." In fact, the village of Barneveld and environs had been slammed by an F5 tornado, 1,200 feet wide, with winds exceeding 200 mph. The town's emergency sirens had not sounded.

In Black Earth, Office Schuyler picked up Dick Ehlert's radio plea. Moments later, Schuyler heard a sound "like twelve freight trains." Suddenly the front of his patrol car heaved into the air. The car skidded backward, propelled by an immense, invisible force. He struggled to regain control of the car, but it kept flying backward for hundreds of feet and careened into a ditch. He sat there, stunned and rubbing the bump on his head as the wind growled around him.

At the Aschliman's, Susan and Charles frantically searched for two-year-old Matthew. They found him on the ground outside, covered with debris. He was badly hurt. Soon he was dead. Moments before the tornado hit, his cry had saved their lives.

In the ruins of Mary Ann Myers's duplex, she quivered and sobbed for her lost loved ones. "I really thought they were gone. I thought: What am I going to do? I'd lost my husband to cancer several years before. I thought, 'I can't handle this.' "

Then she heard their voices. They had survived. The wind had blown Anita, Joey, Jill, and her mother-in-law into the yard. Tiny Joey was seriously injured, but alive. Later they would find his crib, smashed flat. He'd have died if Anita had left him in the crib instead of taking him to bed.

Twenty-five miles away in Madison, reporter Ron Seely of the *Wisconsin State Journal* slept in his Madison home. His phone rang. It was his night city editor, who said something big had happened in Barneveld: A tornado might have hit one or two farms. Seely should check it out. Seely's wife didn't like his leaving in the middle of the night. But that's newspaper life, so he dressed and headed to the car. "It was ugly out. It looked like a tornado night . . . low, rolling clouds." Seely and the newspaper photographer, Roger Turner, rendezvoused and drove in one car toward Barneveld. "We drove over the crest of this hill," Seely recalled, "and when we looked over there, there was nothing but the lights of emergency vehicles, from one end of the ridge to the other. We thought this was a heck of a lot worse than anything we expected."

Back in Barneveld, Doug Salisbury and his wife, Patti, and their children huddled in their flooded basement. Chilled, they wrapped themselves in building insulation that had fallen from the walls. Electrical wires snaked through the water. Doug heard a hissing sound—from gas tanks? He crawled from the basement and hollered into the night for help. He looked toward Sylvia Circle and saw flashlights bobbing in the darkness.

On Jenniton Avenue, Jo Ellen Uptegraw frantically looked for her husband. "Roger," she moaned, "where are you?" Came the muffled reply: "Underneath you." She looked down: She was standing on a fallen door, and Roger was beneath it, uninjured but befuddled. She checked on their eighteen-month-old son, Eric. He was safe in his crib. The crib tilted on its side. Its leg had been severed by a hedge trimmer that had flown through the window.

By candlelight and flashlight, survivors wiped blood from their bodies and scrutinized each other's scars. Teeth were knocked out, skulls cracked, limbs broken. People scratched their skin; fragments of glass fell out. Ruth Ehlert saw her son Keith and tried to hug him, but he backed away: "Don't touch me!" he barked. "There are pieces of glass all over me."

Outside Barneveld, newspapermen Seely and Turner parked and walked across a field into the village. Seely was puzzled by huge, twisted steel balls in the field. Later he learned they were farm machinery. "Then we got to Main Street in Barneveld, and there was just *nothing*—all you could see was foundations and debris. It was a physical shock to see it."

Smashed glass glistened in the streets. Bedsheets dangled from trees. A fork protruded from the brick wall of the firehouse. Dead songbirds lay on the ground, completely stripped of their feathers. Old Christmas decorations sparkled in the road.

A teenager, Mike Holland, clambered from his basement, followed by his mom. He wore only shorts; she, a nightgown. He thought, "Where's my dad?" His father, Harold, was a track coach at Mike's high school. "I asked if anyone had seen him. . . . As I was walking around, I heard someone up yonder, saying, 'Leave him, he's dead.' And I knew for sure it was my dad." It was. Harold Holland was forty years old.

Eight-year-old Cassandra Simon was unconscious. Her parents were dead. Bob Guck, a resident of nearby Avoca whose truck had been waylaid by the storm, gave Cassandra mouth-to-mouth resuscitation. Later she died.

The Lutheran pastor, Reverend Robert Twiton, walked in disbelief through the ruins. He saw young Eric Arneson, a farmer's son. Eric wore a rain slicker and had a lost look in his eyes. "My dad's dead," Eric gasped. "What do I do with my cows? What do I do with my cows?"

Seely helped an old man who was searching through debris for his eyeglasses. They found them—cracked. One man told Seely about riding his bed down the collapsed wall of his house as if it were a ski slope. Seely saw a cat struggling beneath a fallen tree, "still alive and obviously in pain. As I stood there wondering what to do, a police officer came up and unfastened his holster. I walked away and seconds later heard the sharp report of the gunshot."

Through the rainy night, search teams prowled the streets, yelling and waving flashlights as they looked for survivors. At dawn, Seely recalled, "Everybody stopped and looked around, and nobody said a thing." The county coroner "pointed to where the bodies had been discovered. There and there and there, he said." The toll: 9 dead, 200 injured, some 160 buildings leveled—90 percent of the town. The dead were dairy farmer Robert P. Arneson, 52; dairy herdsman Ralph "Rick" Hammerly Jr., 38; high school track coach Harold Kirk Holland, 40; Elaine Slewitzke, 59, a retired U.S. Army sergeant major turned federal farm official; James Slewitzke, 57, a World War II veteran and retired school custodian; Jill Marie Simon, 31, Girl Scout troop leader and mother of three; Vietnam veteran and ambulance driver Bruce Michael Simon, 35; Cassandra Simon, almost 9; and Matthew Aschliman, a month shy of his third birthday.

The village looked ashen, as if drained of color. In the distance, a farmer on a tractor plowed fields that, to Seely, looked unbelievably green.

"The most powerful tornado ever recorded in Wisconsin in the past 140 years" had flattened Barneveld, said meteorology professor Charles E. Anderson of the University of Wisconsin. The Barneveld funnel carved a 36-mile path across the state, passing within 10 miles of 200,000-population Madison. It was the worst of 26 tornadoes in a multistate outbreak.

Tornadoes are famous for their "prankish" behavior. Barneveld was no exception. The tornado crushed almost everything in its path except the most conspicuous object—the water tower, which stood apparently undamaged. (Later investigators discovered a rag stuck *under* the water tower. They concluded that during the storm, the tower had bent slightly and debris flew under its base.) The tornado also destroyed or sucked away certain objects while leaving adjacent ones untouched. The village clerk, Pat Messinger, told me how the tornado sucked heavy maps on her wall out the window, yet ignored bank checks on a nearby desk.

The featherless songbirds reaffirmed an old legend—that tornadoes tear off birds' wings. Over the last century or so, countless farmers claimed that tornadoes plucked their chickens clean. Some scholars were skeptical. But the phenomenon seemed so common that in 1979, researchers Joseph G. Galway and Joseph T. Schaefer of the National Severe Storms Forecast Center (NSSFC) wrote in *Weatherwise* magazine: "While it is not the mission of the . . . [NSSFC] to record tornadoes which deplumed fowls, enough events of this phenomenon have been documented over the past one hundred and forty years to warrant its acceptance."

But the Barneveld tornado definitely shattered one myth: that Blue Mound protected the village. In all fairness, Barnevelders weren't the only people to swallow such a myth, nor will they be the last. Consider the history of Mineral Point, a town near Barneveld. In 1878, a University of Wisconsin professor investigated a tornado that missed Mineral Point. He concluded that a local hill had repelled the funnel, saving the town. But the hill's repellent powers obviously weren't working on the night of June 8, 1984, when an F2 tornado— one of the twisters in the Barneveld outbreak—struck Mineral Point, uprooting twenty trees and damaging several buildings.

Likewise, in Black Earth "there was this myth around town that Black Earth couldn't be hit," Officer Schuyler said. "We're surrounded by hills, and I guess the feeling was that the hills would always protect us from any major storm. Well, that's not the feeling anymore."

Likewise, the inhabitants of Topeka, Kansas, assumed that a local hill, Burnett's Mound, protected them against twisters—until June 8, 1966. On that day, a tornado ripped through downtown Topeka. Seventeen dead, 500 injured. *Time* magazine said: "The legend of Burnett's Mound disappeared into the funnel. 'I never did think it was true,' said a tearful victim. 'But I sure wanted it to be.' "

> *When people have been through [a tornado], the psychological effect is severe; their lives are altered forever.*
> —Ken Wilk of NSSL (1982)

A tornado injures the mind as well as the body. Humans rarely encounter anything else so horrific, short of a wartime blitzkrieg. Psychological damage is inevitable, at least in the short term. The psychic pain is worsened by the twister's creepier pranks. Several weeks after the Barneveld tornado, one resident noticed a curious swelling on his neck. It turned out that an inch-long fragment of glass, probably from a lightbulb, was embedded under his skin. The tornado had turned blades of grass and other debris into miniature bullets, which people plucked out of their flesh for days or months afterward. Later that year, one man coughed up a handful of sand.

After a tornado, a common fear is: Will another one hit soon? Through mid-June, Wisconsin police checked out numerous reports of funnel clouds and twister touchdowns. People fled to their basements at the first rumble of thunder. Civil defense sirens wailed for days. Many "funnel" reports were probably mistakes. Still, the likelihood of a second "hit," while extremely small, is not infinitesimal. For example, in 1948, Tinker Air Force Base in Oklahoma was severely damaged by two tornadoes, five days apart. A church in Guy, Arkansas, has been hit by three separate tornadoes.

A month after the tragedy in Barneveld, "passing storms still cause children to cry and seek the comfort of their parents," the *State Journal* reported. "Many adults who were trapped in their basements complain of being uncomfortable when left alone in a room." Years would pass before they ceased watching dark, wet skies with anxiety; and some have never stopped doing so.

Life was hardest for Barneveld's elderly, especially the men. They and their fathers and grandfathers had built the town with their bare hands; now they viewed its ruins. More than one old, stoic Norwegian face publicly dissolved in tears. "Our older men died within a few years," Pat Messinger told me.

She is a pleasant, thin, grandmotherly woman. We met at the rebuilt Lutheran church, where she and other women were preparing food for a church dinner. "Our old carpenters couldn't handle it," she recalled. "All these houses they had built, all these buildings they had worked on, were gone . . . and they just fell apart—not mentally as much as physically. A lot of our gals in here," she said, glancing toward the happy bustle in the church kitchen, "are widows from the tornado."

A sympathetic world beat a path to Barneveld. President Reagan declared the community a disaster area, eligible for federal aid from the Small Business Administration (SBA), the Federal Emergency Management Agency (FEMA), and other agencies. The Salvation Army, Red Cross, and other private charities came to town, bringing food, coffee, emergency supplies, and good cheer. The homeless were resettled in a temporary trailer park. Neighboring communities pitched in. At a high school in nearby Dodgeville, the public left a "mountain of toys" in the middle of the basketball court, and clothes and bedding on the bleachers.

The news media descended on the razed community. A *People* magazine correspondent said Barneveld "resembled nothing so much as a scene from *The Day After*," a then-controversial TV movie about nuclear war. An insurance man, whose family had lived in Barneveld for many years, cried when he handed Dick Ehlert the insurance check for their home.

Within days, a strange thing happened: Hundreds of bearded, plainly dressed men and humbly garbed women drove into Barneveld, uninvited. They asked: "What can we do to help?" They were Mennonites and Amish, some from hundreds of miles away. They helped Barnevelders rebuild homes and barns, fixed broken plumbing, repaired roofs and walls, and sometimes crouched for hours, picking up debris under the hot sun. "If we . . . didn't come," one asked, "what kind of people would we be?"

After disasters, survivors sometimes turn strangely giddy, as if ecstatic to be alive. The disaster may bring out the best in people. Even loners and misfits may pitch in to help rebuild a town or to assist at soup kitchens. They may suddenly perceive themselves as parts of a warm, loving human family rather than as minor cogs in an impersonal societal machine. Psychologists and sociologists who study disasters have called this new mood "the post-disaster utopia." People start to loosen up, to banter and befriend others whom they would normally ignore. They even start to party (sometimes quite literally), as if saying: "I refuse to be crushed. I refuse to play dead."

In Barneveld, one resident turned her home into a makeshift tavern and placed a painted sign outside: "RUBY'S PUB—HAPPY HOUR EVERY HOUR SINCE JUNE 8." Locals were delighted to cooperate when a beer company came to town and made a commercial about the town's recovery from the disaster (although a few questioned the ethics of exploiting their tragedy to sell beer). "There were times in those days when the greatest need of all wasn't for clothing or food or shelter, but for other people—the clasp of hands, the sharing of tears, the simple pleasure of talk over coffee," Seely later wrote. "Down at the cavernous town hall, at all hours of the day, the women of Barneveld and of many neighboring villages were serving up hot food from big warming pans. There was soup, stew, lasagna, spaghetti . . . There was always plenty of coffee."

Something similar happened after the Topeka tornado of 1966, which killed 17, injured 500, and left 1,600 homeless. The disaster had peculiar, but sometimes welcome, psychological impact, the social scientists James B. Taylor, Louis A. Zurcher, and William H. Key observed in their 1970 study, *Tornado*. "I was excited, and I felt elated—the elation of being alive," said one woman who had lost her home. ". . . You know, it even got to the point where we were joking around a little bit, just like three girls sitting and having an evening together eating sandwiches and having a drink while we watched TV." Psychiatric counselors expressed surprise at some victims' rapid recovery: "I wasn't hearing much about the tornado after two or three appointments," one counselor said. "You'd think they'd talk about it for weeks, but they didn't." The tornado even "cured" one woman's depression by giving her a new purpose in life—serving as a Red Cross volunteer. An observer remarked: "She seemed to find herself through the tornado."

In Barneveld, the "utopian" period manifested itself as a desire to rebuild—not just to rebuild but to rebuild *bigger* and *better*. "The first thing people did was to hurry up and build another house, and some of them built them too big and later wished they hadn't," Messinger said. "Being the village treasurer, I hear this when they come in to pay their taxes: 'Why in the world, when our kids were almost grown, did we build this big house that we're paying for now?' "

The "utopian" mood could not last. Eventually the news media and beer companies go away, and the survivors are left with their wrecked lives, their injuries, their distraught loved ones, their sulking children who suddenly disobey every order. They fill out endless application forms for emergency loans, then wait months for a response. Some grow bitter.

By early autumn Barnevelders routinely complained about delays in getting their Small Business Administration loans.

Winter was coming, and people were sick of their emergency trailer homes. A gas station owner—who had lost his station but still fixed cars in the cold air until his fingers numbed—attended a tense community meeting and snapped that SBA delays were "a bunch of bullshit." A local businessman grumbled: "The real disaster in this community is just beginning."

Couples fought as never before. The University of Wisconsin sent young female psychiatric counselors, who set up a makeshift "office" in a grassy lot. Those brave enough to come for counseling spoke openly about their anxieties. Counselors asked children to express their feelings by drawing pictures of the disaster. It was a good sign, psychologically speaking, if a child drew himself or herself larger than the twister. Some residents developed full-blown depression. Some were institutionalized for stress. Some contemplated suicide.

"I'm tired and depressed much of the time, and that often leads to squabbling," Ruth Ehlert told the nation in a 1986 article for *Redbook* magazine. Others kept their pain private, maintaining a stoic facade while they seethed within. "Being a small town, people convinced themselves they were strong farm stock," Seely said. "They told themselves, 'We can deal with this ourselves, we don't need help.' It was a solid Midwestern community where people are so self-reliant that they think it's a sign of weakness to talk to a psychiatrist or a counselor." As a result, "lots of people suffered mentally. Lots of people still have recurring nightmares. . . . It was just like that: You turn off Johnny Carson, and your house is blown away."

The tornado shook some peoples' religious faith. On the second anniversary of the disaster, the Reverend Frederick Trost told his congregation: "You have shed many tears, sighed many sighs. You have wondered, questioned, doubted, and, I suspect, within many of your souls you have raged." One observer penned a poem:

> Amid the mass confusion
> People asked, "How could this be?"
> "How could He turn His back on us?"
> "When disaster struck, where was He?"

Green Lake is a town about 75 miles northeast of Barneveld. In the second half of 1984, drivers in Green Lake began to have an unusual number of flat tires. They found nails sticking from their tires.

Meanwhile, surprising letters and packages began arriving at the Barneveld post office. The letters and packages contained canceled checks that bore the names of Barneveld residents, photos of Barneveld residents, records from the Barneveld bank, and other Barneveld-related items. People in towns many miles away (some on the other side of the state) had found the items in their yards. The debris had fallen from the sky, from the winds of the same storm that caused the Barneveld tornado. The debris included countless nails, perhaps ripped from the walls of someone's home.

Word of the far-flung debris reached Professor Charles E. Anderson at the University of Wisconsin in Madison. Anderson was a sixty-five-year-old meteorologist and professor of space science and engineering. He had received his Ph.D. in meteorology at MIT, becoming the first African American to win a doctorate in that subject. During World War II he was a captain in the U.S. Army Air Force and a weather forecaster for the legendary all-black Tuskegee Airmen Regiment.

Anderson ran newspaper ads asking the public to report mysterious debris. The ads triggered more than 200 tips from across the state. The tornado had blown away a recreational vehicle; its bumper and license plate were found in Lodi, almost 30 miles from Barneveld. The tornado removed a driver's license from a man's wallet and hurled it to Appleton on the other side of the state. (The wallet remained in Barneveld.)

The Barneveld tornado "sucked up tens of tons of material from the village . . . [that] fell out along a trail from Barneveld to north of Green Bay, some 137 miles away," Anderson later reported at a severe storms conference. The "heavy" material (defined as "objects weighing one pound or more, such as pieces of plywood, aluminum siding, boards, etc., from buildings") fell in three distinct paths, all within a narrow, 85-mile region from Barneveld to Redgranite. "Light" debris (defined as items weighing less than a pound, "such as tar paper, shingles, soffit [ceiling material], insulation bits . . . and occasionally bundles of light materials such as bank check statements in envelopes") dispersed more widely. In Mayville, 86 miles to the northeast, someone found a bank pouch marked "Barneveld State Bank." Anderson traced some items' "trajectory" through the storm, based on their mass, aerodynamics, and other factors. He concluded that the bank pouch had, incredibly, soared all the way to the top of the thunderstorm—to its icy roof of cirrus clouds—then plummeted to Earth.

Anderson's Barneveld study inspired a new branch of tornado science: long-distance debris dispersal, which was later an important part of the VORTEX project.* Debris dispersal isn't merely of theoretical interest; it has important practical implications, as shown by Anderson's successor, John T. Snow of the University of Oklahoma.

An amiable, burly man with a moustache, Snow left Purdue University a few years ago and now heads the College of Geosciences at the University of Oklahoma. He runs VORTEX's debris dispersal study project, which has made some surprising and unsettling findings. One finding is that long-

---

*

The subject itself was old news, however. For centuries, laypeople had told stories about tornadoes that blew objects tens or hundreds of miles. The strangest examples involved "falling" frogs and fish. The ancient Greek scholar Athenaeus wrote: "I know, too, that it has rained fishes in many places . . . in Chersonesus it rained fishes for three whole days . . . certain persons have in many places seen it rain fishes, and the same thing often happens with tadpoles." In 1949, Louisiana researcher A. D. Bajkov wrote *Science* magazine about a "rainfall of fish": "I was in the restaurant with my wife having breakfast, when the waitress informed us that fish were falling from the sky. . . . There were spots on Main Street . . . averaging one fish per square yard. Automobiles and trucks were running over them. Fish also fell on the roofs of houses."

distance debris dispersal is far more common than previously thought. Also, it appears that even "moderate" tornadoes can transmit debris great distances.

In 1994 and 1995, Snow and his students followed Anderson's example by running ads asking people to report mysterious debris. They also opened a toll-free telephone line. After tips started to come in, Snow sent his graduate students out in the field to recover debris after tornadoes. They had to act quickly before cleanup crews hauled the debris to garbage sites or landfills. During the Friona tornado, for example, the vortex hit a building that contained private financial records. One such record was a canceled check. The tornado sucked up the check and blew it all the way to Amarillo, 65 miles to the northeast. Other Friona and Dimmitt detritus rained across north Texas. Many items fell tens of miles from their points of origin. They included shingles, building insulation, a 5.25-inch computer disk, a piece of wood from an airplane, tin sheets, Styrofoam, a cassette tape, and silk flowers. The flowers may have come from a cemetery near Friona. The students "come back with a pickup truck full of sheet metal and paper," Snow recalled. "We have a deed to somebody's home, and a deed to a feed mill. It came out of the sky!" He smiled at the oddity of it all.

Then they sifted through debris and traced much of it back to its place of origin. For example, they traced photos by sending copies to people in communities hit by tornadoes. An accompanying letter asked: Do you recognize anyone in this photo? "People have said, 'Oh yes, that's my granddaughter' and 'That's my high school picture.' " His students traced the origins of debris as diverse as a golf course flag, including part of the plastic pole (which flew 43 miles), a telephone directory (56 miles), and fiberglass roof fragments (about 31 miles). They also traced a man's jacket back to its owner, 40 miles from where it landed. "It had his name monogrammed on it."

Most tornado debris falls to the left of the storm path. Why? Presumably the debris rises into the storm and is swept around by the mesocyclone, and some is ejected toward the north (left). A minority of debris falls to the right, however.

Based on the historical record, researchers have always assumed that long-distance debris dispersal was rare. Only the strongest tornadoes would blow debris for tens or hundreds of miles, they thought. Grazulis reviewed 13,000 tornado reports from 1871 to 1990 and found only 121 cases with evidence that debris traveled more than 5 miles. Tornadoes have blown chickens and a carton of deer hides 6 miles, a cow and an airplane wing 10 miles, a jar of pickles 18 miles, dead ducks 25 miles, a music box 35 miles, a necktie rack 40 miles, and a wedding gown 50 miles. The record distance is 210 miles, traveled by a canceled check during the Great Bend, Kansas, twister of 1915. This storm destroyed 160 homes and killed 2 people and 1,000 sheep; "hundreds of dead ducks fell from the sky," Grazulis notes.

But Snow's research indicates that long-distance dispersal is remarkably common. During VORTEX, Snow's team collected "reports from 17 separate [tornadoes] . . . over 2 years." That rate hints that long-distance dispersal is "almost 12 times more [common] than what the historical record would suggest."

More important, the Snow team realized that even "moderate" tornadoes can blow debris great distances. An F2 twister struck Moberly, Missouri, on July 4, 1995. It hurled unopened cans of soda 87 miles and a brick (which weighed several pounds) roughly 28 miles.

These findings raise a scary question: What would happen if a tornado struck a chemical plant, pesticide factory, or other toxic facility? Might the high winds scatter toxins for hundreds of miles? Might authorities be forced to evacuate communities, perhaps even entire regions, to protect people from the deadly whirlwind of poisons?

The risk of such a catastrophe may be worse than is generally realized, for two reasons. First, F2-or-stronger tornadoes are by no means uncommon. According to the National Severe Storms Forecast Center in Kansas City, one third of all tornadoes rank F2 or higher in intensity. That works out to several hundred *known* F2-and-worse twisters annually in the United States. (The actual number is probably much higher. Experts estimate that only one in three tornadoes is seen and recorded.) Some of these storms pass near industrial facilities.

For example, a March 1995 report by the U.S. Department of Energy, which manages the nation's extensive complex of nuclear weapons manufacturing plants, stated that since 1953 at least nine tornadoes have occurred "on or near" the network's Savannah River Site in South Carolina. At Savannah River and other complex sites, large amounts of radioactive waste have accumulated in equipment and terrain (due to leakage) over several decades. So far no tornado has hit a radioactive waste site. But the risk, while small, may not be negligible. A 1993 report by the National Academy of Sciences, *Wind and the Built Environment,* complains that U.S. Department of Energy and Department of Defense facilities "often suffer significant damage from extreme wind events. Nonetheless, their support for the wind-engineering and wind-hazard research aimed at reducing these losses is almost nonexistent."

Second, the potential "targets" are everywhere: Our high-tech society relies on numerous deadly substances (toxic chemicals, explosives, bacterial specimens, radioactive materials, etc.) for scientific, commercial, and military purposes. Snow is especially concerned about radioactive substances from nuclear materials manufacturing plants; medical waste such as infected, bloodstained rubber gloves; large stocks of agricultural chemicals; and contaminated soil at "Superfund" cleanup sites. If "long-distance transport of debris by tornadic thunderstorms is a common event, then tornadoes pose serious [environmental] hazards well beyond the direct impact of their high winds," he and his colleagues wrote in a recent paper.

Snow's research may help scientists improve computer simulations of airflow within tornadic storms. Using these computer models, scientists could test their theories of storm structure by forecasting where debris will fall after a particular tornado, then comparing the forecast with the actual locations of "fallout."

Such futuristic computer models might save both lives and ecosystems. Suppose that a tornado hit a chemical factory and blew its toxins across the countryside. Emergency teams could use a computer simulation to project where the poisons fell and to decide whether to evacuate any communities. They could also use the model to plan the fastest and most efficient way to clean up the toxins before they caused irreparable damage—say, by seeping into the soil and poisoning the groundwater.

Our society relies more and more on scientific wonders— exotic chemicals, viruses for vaccines, all kinds of biotechnological mutations, radioactivity for cancer treatments and hospital research—a cornucopia of agents that, unleashed into the environment, could transform from friends into foes. In an increasingly high-tech future, the biggest danger may be what the tornado hits, not the tornado itself.

Now, a decade after the devastation, Barneveld is rebuilt, all shiny and new. The population, which fell to 500 or so after the twister, is now higher than ever, around 850. A half-finished housing complex sprouts on the edge of town. The market and gas station are bigger and more modern-looking than ever. The rustic architecture of yesteryear has disappeared, and taken with it a tactile sense of connection to the past, to those deep familial roots that stretch back to Norway and Germany and Ireland and Sweden.

On the edge of town, a small park has a concrete public memorial, where a plaque lists the victims of the tornado. The plaque vaguely acknowledges "the power of people helping people." The memorial is the only structure in town that looks run-down, as if untended—as if most people would rather forget what happened.

Forgetting is hard, though. Ghosts are everywhere. For years after the tornado, Reverend Twiton cried when he recalled one victim—"little eight-year-old Cassie. She was my friend. She used to come over and visit while I was working on the church's garden. She used to come to church and sit all alone in the front row. I'd look down and there would be Cassie."

Mary Ann Myers's grandson, Joey, spent months in therapy learning how to walk. Today he is twelve, takes medication for seizures, and is a "special needs" child with the mental level, for most cognitive tasks, of a three- or four-year-old. She proudly shows a photo of him. He has big, dark eyes and a soft smile. "He's a beautiful little boy," she said. "If he came in that door now, he'd say, 'Hi Grandma, I love you,' and give me a big hug."

The press, once so welcome, is now almost persona non grata. People fume about all the times they were misquoted, or about the hard-boiled reporters from big cities who barged into a memorial service with microphones extended and TV cameras whirring. When this writer asked Anita Jabs for an interview, she was reluctant. "Usually I say no," she said. She finally agreed to talk while ringing up customers' purchases on the cash register at Ron's Store, the well-stocked, cheerfully lit market that she runs with her husband.

After the twister, an economic analyst from the University of Wisconsin cautioned the couple against rebuilding a bigger store, she said. "We were told that we don't have the right numbers of customers.

"But we went ahead with [rebuilding] anyway. . . . We let our hearts get in the way of our good sense. We tried to get things in our own personal business and personal lives put back together as fast as possible—and that's a mistake." Had they taken the time to think about it, "we might not have rebuilt. . . . We don't make a great deal of money for putting in the hours that we put in.

"I guarantee you, if the tornado happened again, I would be out of here. It's too much to go through again."

At one point, a customer came through the grocery line. She whispered something to Anita about reporters. Anita nodded as she stabbed the cash register keys and muttered, "The tornado will *never* go away."

Not in many Barvevelders' minds, anyway. At night when the bad dreams come, it screeches through the darkness again and again and again. In their waking hours, the first drop of rain forces them inside their homes; the first strong gust of wind drives them to their basements.

And on spring or summer evenings, when thunderheads jam the horizon and sparkle with lightning, Mary Ann Myers no longer sits outside to enjoy the show. She stays indoors.

CHAPTER 3

American Gothic

Most scientific problems are far better understood by studying their history than their logic.
—Ernst Mayr, *Growth of Biological Thought*
(1982)

*A*mericans' battle against tornadoes began more than a century ago. Settlers of the American West encountered terrifying twisters, so terrifying that a few feared the West would prove uninhabitable. However, the modern age of tornado science didn't begin until the 1940s and 1950s, when twisters began to annoy a particularly prickly target: the U.S. military, which launched a tornado-forecasting project. The U.S. Weather Bureau followed suit as the nation urbanized and tornadoes began to clobber fair-sized towns.

*T*he men, women, and children who settled North America gasped at its physical grandeur: its sprawling mountain ranges, vast gorges, and thousand-mile rivers. Everything about the continent seemed bigger than life—and its weather was no exception. Early pioneers were occasionally amazed to see long swaths of forests stripped of trees. A resident of Guelph, Ontario, Canada, in the 1820s noted "large gaps or lanes" that "had the appearance of a wide road slashed through the forest . . . [where] the trees appear to have been twisted off at the stumps, or turned up at the roots, as if some monster of infinite strength had passed that way. . . ."

The late-nineteenth-century British meteorologist Douglas Archibald referred to "the terrible tornadoes of America which have been known to carry solid objects like wooden church spires a distance of 15 miles and kill hundreds of people in the space of a few minutes. . . . The destruction caused by these tornadoes in America is hardly realized in Europe which is so happily exempt from them."

The earliest colonial record of a possible tornado dates from July 5, 1643. The Massachusetts governor noted in his diary that a "sudden gust" had wrecked a meeting house and blown over a tree, killing a Native American. However, Thomas P. Grazulis, a leading tornado historian who did climatological research under contract to the Nuclear Regulatory Commission, suspects the 1643 "tornado" was actually "a gust front or downburst-type storm." It's an early example of a persistent historical puzzle: When is a "tornado" not a tornado? Our ancestors (particularly early newspaper writers) tended to use the words "tornado," "waterspout," "cyclone," and "hurricane" interchangeably.

The "first confirmed true tornado" struck on July 8, 1680, at Cambridge, Massachusetts, Grazulis says. The funnel killed a servant and unroofed a barn. Two years later a twister ravaged a forest near New Haven, Connecticut. "Great limbs were carried like feathers," a chronicler wrote. Over the next century (which climaxed with the American Revolution), only 17 tornadoes were recorded. The scarcity of eighteenth-century tornado reports reflects the demography of those times:

The colonial population was small and restricted largely to the northern part of the Atlantic coast. Few tornadoes were seen because there were few witnesses to see them.

In the next century, white settlers flooded into the nation's interior. They crossed over the Appalachian Mountains and settled in the lush croplands of the Mississippi River, or on the windswept plains teeming with buffalo. There they encountered many hardships: starvation, sickness, bloody conflicts with the natives. But their harshest foe was the sky. Blizzards entombed cattle and cattlemen under snowdrifts; lightning sparked vast prairie fires; drought desiccated a farmer's crops, then hail smashed them flat.

But the tornadoes . . . well, those exceeded the purplest prose. In June 1814, the naturalist John James Audubon saw a tornado near Shawneetown, Illinois.

> Two minutes had scarcely elapsed [since observing a yellowish oval spot toward the southwest] when the whole forest before me was in fearful motion. I saw, to my great astonishment, that the noblest trees of the forest bent their lofty heads for a while, and, unable to stand against the blast, were falling into pieces. . . . [Tree branches] whirled onward like a cloud of feathers. The horrible noise resembled that of the great cataracts of Niagara, and it howled along in the track of the desolating tempest.

Native Americans had settled the continent many millennia earlier. They were no strangers to atmospheric vortices ranging from dust devils to tornadoes. Barry H. Lopez has recorded a bawdy Arapaho legend about an encounter between Coyote, the archetypal Native American "trickster," and "Whirlwind Woman," who "sometimes crawled along in the shape of a caterpillar."

> "I want you for my sweetheart," he said to her.
> "No," she answered. "I am used to moving all the time. I

do not like to stay in one place. I travel. I would not be the wife for you."

. . . He grabbed her and tried to lay down on top of her.

Whirlwind Woman began spinning and she caught Coyote and threw him headfirst into the river bank.

The Crow tribe told of a boy, Bear White Child, who is picked up by a singing black cloud, then returned to the ground. In prehistoric Kansas, a twister ravaged the Potawatomi tribe. Survivors buried the dead on a hill later named Burnett's Mound, southwest of the future city of Topeka. The natives recounted the disaster in a song that lamented: "The grass is moving, the trees are moving, the whole Earth is moving."

A few white settlers were enchanted by the whirring funnels. "A sublime sight," remarked Father Pierre Jean de Smet, a priest who accompanied settlers from Indiana to California a few years after the presidency of Andrew Jackson. He wrote:

> A spiral abyss seemed to suddenly be formed in the air. The clouds followed each other into it with such velocity, that they attracted all objects around them. The noise we heard in the air was like that of a tempest. . . .
>
> The column [funnel] appeared to measure a mile in height; and such was the violence of the winds which came down in a perpendicular direction, that in the twinkling of an eye the trees were torn and uprooted, and their boughs scattered in every direction. But what is violent does not last. After a few minutes, the frightful visitation ceased. . . . Soon after the sun reappeared: all was calm and we pursued our journey.

The enchantment didn't last, either. On May 7, 1840, a tornado massacred 317 people in Natchez, Mississippi. Residents were eating lunch when skies darkened "so as to require the lighting of candles," one newspaper said. Next rain began falling in "tremendous cataracts." Then the tornado hit. The Natchez *Weekly Courier and Journal* reported:

HORRIBLE STORM!!
NATCHEZ IN RUINS!!!

Our devoted city is in ruins, and we have not a heart of stone to detail while the dead remain unburied and the wounded groan for help. Yesterday, at 1 o'clock, while all was peace, and most of our population were at the dining table, a storm burst upon our city and raged for half an hour with most destructive and dreadful power. We look around and see Natchez, yesterday lovely and cheerful Natchez, in ruins and hundreds of our citizens without a shelter or a pillow. Genius cannot imagine, poetry cannot fill up a picture that would match the ruins and distress that every where meets the eye.

. . . [We see] a scene of desolation and ruin which sickens the heart and beggars description—all, all is swept away, and beneath the ruins still lay crushed the bodies of many strangers. It would fill volumes to depict the many escapes and heart-rending scenes. . . . Mrs. Alexander . . . was found greatly injured with two children in her arms and they both dead! The destruction of flat boats is immense; at least sixty were tossed for a moment on a raging river and then sunk, drowning most of their crews. . . . No calculation can be made of the amount of money and produce swallowed up by the river.

Tornadoes' fury dazzled the nineteenth-century mind. The scientist Elias Loomis investigated a tornado that blew through Ohio in 1842. The twister fired boards up to 18 inches into the soil. He simulated the impact by inserting similar boards into a six-pounder gun and launching them into a hillside. On this basis, he calculated that the tornado's winds were 682 mph—close to the speed of sound. Other nineteenth-century theorists suggested tornado winds exceeded 2,000 mph. And this was an era when people gasped at the sight of trains moving a few tens of miles per hour! It is no wonder that some nineteenth-century scientists regarded tornadoes as a *mysterium tremendum,* a wrench thrown into the shiny machinery of Newtonian physics.

After the 1896 tornado in St. Louis, Willis L. Moore, chief of the U.S. Weather Bureau, visited the scene and found a narrow piece of wood "shot through five-eighths [inch] of solid iron on the Eads Bridge . . . [like] shooting a candle through a board."

Westward immigration exploded after the Civil War, when the transcontinental railroad opened. Dee Brown's *Hear That Lonesome Whistle Blow* (1977) says "tornadoes could lift a train off the track. One of the legends of the Kansas Pacific [railroad] concerns a tornadic waterspout that dropped out of a massive thunderstorm, washed out six thousand feet of track, and swallowed up a freight train. 'Although great efforts were made to find it,' said Charles B. George, a veteran railroad man, 'not a trace of it has ever been discovered.' "

Frontier life was dawn-to-dusk toil, and every ambulatory member of a pioneer family had to pitch in by pumping the well, milking the cows, and ploughing the fields with stubborn mules. Twisters injected sudden fright into their drab lives. Usually the tornado was just a dark, harmless ribbon on the horizon. Joanna L. Stratton's *Pioneer Women* (1981) quotes a settler who recalled a hair-raising school day in frontier Kansas:

> At the recess period they . . . saw a well-defined funnel-shaped cloud . . . move rapidly in their direction. They were almost panic stricken when they saw it so near that it obscured from their view an old sod house that stood two miles to the southeast.
>
> There was a cyclone cellar under the school house, but no door had yet been made leading to it. They all ran into the house, where the teacher seized the kindling hatchet with the idea of chopping a hole in the floor or prying up some boards, when one of the white-faced children cried out from the doorway, "It's done turned, teacher. It's going straight north." And sure enough, in that freakish way that tornadoes have, it shifted its course.
>
> The people in the district never grew tired of teasing the teacher about scaring away a tornado with a hatchet. . . .

Everett Dick's *Tales of the Frontier* (1963) tells about . . .

> . . . the afternoon of May 30, 1879, [when] Gerhart Krone, a homesteader living near Delphas, Ottawa County, Kansas,

came in from his fields when a gentle rain was succeeded by a fusillade of hail. About three o'clock, Krone looked across the plains to the southwest beyond the Solomon River, and saw an angry, greenish-black cloud from which hung a giant, swaying funnel. It was moving toward the northeast, and as it approached, a dark, smoke-like mass of flying debris was plainly visible at its base.

. . . He himself was running toward the northeast, with his grown daughter beside him, when the twister struck the house, reducing it to kindling wood. Krone was flung to the ground, then lifted up and dropped down again several times, sustaining severe injuries. His daughter, after being carried a distance of two hundred yards, was hurled against a barbed wire fence and killed instantly. Every shred of clothing was ripped from her body, and she was covered with black mud. . . . Another daughter, Anna, had a piece of board driven nearly through one thigh; then the wind itself extracted the board, leaving a seven-inch gash. The doctor later found nails, straw, and splinters of wood driven into the ghastly wound.

Grotesqueries abounded. "The head of an infant, and the arms and legs of a grown person, were brought from many miles to the westward" by an 1860 Iowa twister. In Wisconsin in 1865, "tree tops were filled with feather beds, chairs, and clothing; all kinds of livestock were either dead or writhing on the points of the broken branches." In Georgia in 1875, a man was sitting in an outhouse when a twister hit, driving pieces of bone into his brain. "These were extracted by Drs. Branham, Hendrick, and Yancey, and a tablespoon of brains taken out," a newspaper noted. "The man is doing well." Other accounts tell of humans stripped naked and a cat "crushed as flat as if passed through a cider press"; of colts thrown 1,000 feet into the sky; of a herd of 160 crazed cattle that stampeded into a funnel, where 155 had their necks broken; of human "eyes, ears, mouths and wounds [that] were filled with mud, chaff, straw and bits of wood. Two weeks of washing could not clean the skin." A Kansas woman's lifeless body was found buried headfirst up to her shoulders in the dirt. Such events lent a Gothic quality to frontier life.

The Wizard of Oz exaggerated: Tornadoes usually lofted peoples' homes intact by only a few feet. A few houses flew much higher, but most disintegrated in the ascent. In 1875 in Georgia, John Steap stood in the doorway of his house as it sailed 75 feet on the breath of a tornado. En route he felt as if "it was gliding over a pond of water . . . he felt no shock when it was put down on the earth again." An 1878 Iowa twister raised a house "like a toy, twisted it about twenty feet into the air, and dropped it 300 feet from where it stood. Not enough timber can be seen to build a chicken coop."

"Stereoscopic photos" of tornado damage were a Victorian novelty. Stereoscopic photos are near-twin images of the same scene that, when viewed through special eyeglasses, acquire a "3-D" quality. They were a popular form of Victorian home entertainment. Hundreds of different stereoscopic images of tornado wreckage were mass-reproduced and sold. They were the "tornado videos" of their day.

June 12, 1899, was circus day in New Richmond, Wisconsin. The day was hot and humid, and the town was packed with families and revellers. Around 6:00 P.M., residents observed a "top-shaped cloud" over Lake St. Croix. Then a twister touched down. Terrified locals sought shelter in the basement of a dry goods store, but its brick structure was "sucked up by the tornado and hurled back down on the crowd in the cellar," according to a modern chronicler, Mary Sather. A fire broke out in one building, where a man's foot was trapped under lumber. He screamed as the inferno neared: "Cut off my foot! Cut off my foot!" And then: "Kill me! Kill me before I burn to death!" The fire enveloped him. One of Ms. Sather's aged relatives fled to a basement and told two women to join him. They hesitated and were decapitated by flying debris. After the storm, witnesses saw the once-thriving community smashed to pieces. The ruins were enveloped in a "weird, uncanny light, green and hideous." The body count was 117.

People living in an age before atomic bombs, terrorist explosions, and 747 crashes struggled to describe the violence they had just witnessed. Nothing in their experience, not even the Civil War, could match the apocalyptic sights of a tornado strike. Their accounts bordered on the biblical: "The earth trembled as with an earthquake. The air was filled with sulphurous fumes. Vast waterspouts gyrated to every point of the compass. A roar like Niagara Falls filled the stricken people with fear. The shrieks of the wounded and the wails of the bereaved blanched the faces of the stoutest."

Hope was coming, however, in the emerging science of meteorology.

*A*ny child who watches a TV weather show knows two facts that early Americans didn't: (1) Storms move horizontally across the Earth, and (2) many storms rotate.

Before the American Revolution, scientists assumed that weather was a local phenomenon. A storm formed in a particular region, then died there. This provincial view of weather was rejected by a man who revolutionized both science and politics.

In Philadelphia on October 21, 1743, Benjamin Franklin looked forward to watching a lunar eclipse. But a storm struck, spoiling the view. Several days later, he read in out-of-town newspapers that residents of Boston (250 miles to the northeast) had seen the eclipse, then endured a violent storm the next day. Was it the same storm that struck Philadelphia on October 21? Franklin concluded it was. Weather moves!

Later scientists realized that American storms typically move from the western half of the compass to the eastern half (usually from southwest to northeast). This discovery suggested a way to forecast the weather: Find out what weather conditions are like to the west.

Unfortunately, the storms moved faster than eighteenth-century communications. By the time Easterners got wind of western storms, the storms had come and gone.

The electric telegraph changed everything. Pioneered by Samuel F. B. Morse and others in the 1830s, electric telegraphy permitted "real time" communication at the speed of light—186,000 miles per second. Train station operators used telegraphs to route trains and prevent collisions. Weather scientists exploited the train network to gather weather information from frontier forts and towns. "I would frequently write upon the bulletin board . . . what and when weather changes were coming," one telegrapher wrote in 1846. "Frequently this was with such accuracy as to create considerable comment and wonder." In 1870, Congress charged the U.S. Army Signal Service with "taking meteorological observations at the military stations in the interior of the continent and at other points in the States and Territories . . . and for giving notice on the northern [Great] lakes and on the seacoast by magnetic telegraph and marine signals, of the approach and force of storms."

Early researchers such as James Pollard Espy realized that storms are caused by the ascent or convection of warm air. At higher altitudes the air expands and cools, so its water vapor condenses into water droplets visible in the form of clouds. Espy even suggested fighting droughts by setting enormous forest fires that would spawn rain clouds!

The second crucial discovery was that many storms *rotate*.

Nowadays every schoolchild knows that air spirals into a tornado. But most scientists didn't recognize this truth until the second half of the nineteenth century. Previously, they thought that tornadoes sucked up air in straight lines rather than in a spiralling fashion. In 1838, William C. Redfield investigated a New England tornado that resembled "the proboscis of an enormous elephant." He walked seven miles of the twister's path and analyzed the directions in which trees fell. He concluded their orientation was "decisive evidence" of winds rotating around the tornado, "in the direction from right to left or which is contrary to the hands of a watch"— that is, counterclockwise or "cyclonically." Redfield's views led to a celebrated debate with Espy, who denied that tornadoes rotate at all.

Nowadays, we know that *almost* all tornadoes and *all* hurricanes in the Northern Hemisphere rotate cyclonically (but anticyclonically, that is, clockwise, in the Southern Hemisphere.) Why? The American scientist William Ferrel attributed their direction of rotation to the rotation of the Earth—specifically, to the Coriolis "force."

The Coriolis effect is named for the nineteenth-century French physicist Gaspard-Gustave de Coriolis. As wind blows, Earth rotates beneath it. This causes the wind to appear to curve slightly. In the Northern Hemisphere the wind appears to curve to its right; in the Southern Hemisphere, to the left. You can mimic this effect with a piece of chalk and a phonograph player: As the turntable spins, draw a straight line from the center of the record toward you. Then stop the record and observe how the line appears to curve to the side. The same effect causes Northern Hemisphere low-pressure cells (sites of rising air) to rotate counterclockwise or cyclonically, and high-pressure cells (sites of sinking air) to rotate clockwise or anticyclonically. The effect reverses in the Southern Hemisphere, where low-pressure cells rotate clockwise and high-pressure cells, counterclockwise.

Ferrel was right about hurricanes—but *not* about tornadoes. Tornadoes are too small for their rotation to result directly from the Coriolis effect. (Another popular myth is that the Coriolis force makes toilets flush clockwise in the Northern Hemisphere and counterclockwise in the Southern Hemisphere!) However, almost all Northern Hemisphere tornadoes rotate counterclockwise anyway because they inherit the counterclockwise motion of their parent storms. (This may seem like splitting hairs. But progress in physics often depends on the precise splitting of theoretical hairs!)

Ferrel correctly explained how a tornado develops its low air pressure. It spins so rapidly that air accelerates outward, preventing air from rushing into the funnel from its sides. Instead, new air penetrates the funnel through its lower tip, as in a vacuum cleaner. This slows the ability of the funnel to replace air sucked up by the updraft, and drives the air pressure even lower. Air pressure inside tornado funnels probably drops by about 10 percent—equal to about a 2-inch drop of a barometer. Similar pressure drops also occur during hurricanes.

The discovery of tornadoes' rotation helped nineteenth-century physicists explain their extraordinary violence—in part, anyway. This partial explanation is based on the "law of the conservation of angular momentum." The best-known example of angular momentum is a spinning skater. Watch Nancy Kerrigan on TV as she spins with her arms outstretched. What happens as she wraps her arms around herself: She spins much faster, right? Why? A scientist would describe Nancy's energy in terms of momentum (her velocity multiplied by her mass). The angular momentum law says her velocity multiplied by the radius of her spin must equal a constant value. Therefore, if she changes her radius, then her rotational velocity must also change to keep the constant "constant."

Likewise, nineteenth-century scientists assumed that large, vertical columns of air such as cyclones start spinning and gradually contract. As they contract, they spin faster and faster until (thanks to the law of the conservation of angular momentum) they're spinning at hundreds of miles per hour. This didn't fully explain the tornado's violence, as we shall see. But it was a step in the right direction.

Is there a probability that, with the closer settlement of the West, tornadoes will become more frequent?
—T. B. Maury, _North American Review_ (1882)

In the 1880s, westward expansion was in full swing. Americans flooded into lands where twisters lurked. The result was hundreds of horrible encounters between settlers and the lethal funnels. Some blamed the tornadoes on human interference with nature; others dismissed the tragedies as happenstance. The resulting dispute was a Victorian microversion of today's "climate change" debate.

In 1883, George Clinton Smith charged in *Popular Science Monthly* that tornado rates were rising because of "radical changes . . . in our atmosphere and climate" caused by "the construction of great railroad-belts across the continent and the erection of a vast network of telegraph and telephone wires." Smith believed the railroad and communications lines upset "the equilibrium of electric forces" in the atmosphere, triggering tornadoes. Other critics blamed the tornado wave on the settlers' destruction of Western forests and over-farming of its soil. These practices, they argued, diminished the amount of water vapor rising into the sky—hence, droughts. The droughts, in turn, triggered tornadoes (by some means not explained).

Nonsense, replied the writer T. B. Maury. "If aridity of soil was sufficient to bring about [more tornadoes, then tornadoes] would abound in the [dryer] trans-Mississippi Plains more than in the immediate Mississippi Valley, and in many rainless regions of the globe in which the tornado properly rarely occurs," he wrote in *North American Review*. "Man is as powerless to work any change which will augment or diminish the number of tornadoes, or to disturb the ponderous atmospheric machinery which produces them, as the puny fly is to retard or accelerate the motion of a powerful steam-engine." As we now know, the real reason for the "increase" in tornadoes was demographic: The more people who moved West, the more "witnesses" there were to see tornadoes that would have otherwise been missed.

Nonetheless, tornadoes threatened to panic Western immigrants. Maury complained about the

wide-spread impression that, with the deforesting and settlement of the West, tornado-visitations have increased, so that a prominent journal recently raised the question whether their frequency and destructiveness will not have "a permanent effect on the settlement and prosperity of the country." We are even told that in some places the alarm created by these storms is so great that "the people are not only digging holes in the ground and building various cyclone-proof retreats, but in many instances persons are preparing to emigrate and abandon the country entirely."

In short, tornadoes threatened Manifest Destiny, and the U.S. Signal Service vowed to fight back. It did so in the person of Sergeant John Park Finley.

Finley was a beefy-looking man with a mustache. He was America's first real "tornado expert." In the 1870s, he inspected twister damage by touring the Western states by horse and buggy. He made meticulous maps of farms that showed the exact paths and impacts of tornadoes. (One Finley map included a curved line that represented the aerial path of a hired man, whom a tornado blew 400 feet.) He found that the worst damage was confined to narrow areas that averaged 1,400 feet wide, and that few twisters traveled more than 20 miles. He marveled at the "most terrific" damage near Irving, Kansas, where a new iron and steel bridge was "completely twisted into shapeless ruin."

Finley also pioneered tornado climatology, the study of tornado frequencies. He enlisted more than 2,400 volunteers for a tornado-observing network. They initially reported an average of 10 to 60 tornadoes per year. In 1880, the annual count passed 100 for the first time. (The present annual average exceeds 1,000!) "Without this man's perseverance, there would probably be no [nineteenth-century tornado] records of value," said tornado climatologist Thomas P. Grazulis.

However, tornado statistics were even trickier in the nineteenth century than they are today. For example, in 1887, Finley drew a map showing the locations of tornadoes in several Midwestern states. The map was thick with dots except in Oklahoma—nowadays known to be a hotbed of twisters! How did Finley err? It wasn't his fault. Back then Oklahoma was "Indian Territory," from which few reports were received. The flawed 1887 map is a stark example of the intimate link between demography and tornado frequencies, a link that still confounds tornado climatologists today.

Finley also began forecasting tornadoes. In 1879, he advised the War Department to post a weather observer in Kansas City during tornado season. The observer could telegraph news of severe storms to points east. "To get the right information to the proper point before the occurrence of the dangerous [tornado] phenomenon, thereby affording opportunity to provide against its ravages, is the great desideratum. It can be done." He began issuing his own forecasts in 1884, based on limited weather data from across the nation. Unfortunately, weather stations were few and far between, especially in the tornado-plagued Midwest.

After several years, Finley's bosses decided that tornado forecasting was a bad idea: It might spark panics. So the Signal Service banned the use of the word "tornado" in forecasts. The U.S. government didn't resume tornado forecasting until the mid-twentieth century. The rest of Finley's program soon withered and died.

Afterward, American tornado research entered a dark age that lasted until the 1940s. The center of tornado research shifted to Europe. There, an important tornado scientist was Alfred Wegener (1880–1930) of Germany. Wegener was a pioneer in more ways than one. He championed "continental drift," the controversial theory that continents drift around the Earth. (The theory is now widely accepted in a revised form called plate tectonics.) He argued that lunar craters were created by meteors, not by volcanic eruptions as was then generally believed. (He tried to prove his case by hurling projectiles at basins of powder, cement, plaster, and mud.)

In 1928, Wegener wrote a paper that anticipated a modern concept of tornadogenesis (the cause of tornadoes).

As noted earlier, a tornado may descend from a much wider column of spinning air, the mesocyclone. To visualize the mesocyclone, imagine a tall cylinder rotating on its vertical axis (like a barber pole). Inside the thunderstorm, this rotating cylinder of air sprouts smaller offspring—thinner cylinders that also rotate on their vertical axes. Some of these smaller cylinders descend to Earth as tornadoes.

But how do mesocyclones themselves form? That is, what causes a miles-wide cylinder of air to rotate around its vertical axis like a barber pole? Wegener offered a simple explanation. First, high-speed upper-air winds sweep over lower, slower air and cause a cylinder of air to rotate *horizontal* to the ground. (As an analogy, move your hand over a pencil on a desk. The pencil rolls across the table, rotating along its horizontal axis.) Next, the cylinder passes over a warm updraft. The updraft shoves the middle of the cylinder upward (imagine using your finger to push up the center of a clothesline). As a result, the cylinder acquires an upside-down "V" shape. Eventually the updraft raises one "arm" of the "V" into a vertical position. (See graphic.) Wegener called this vertical rotating cylinder the "mothervortex," or *mutterwirbl* in German. We call it the mesocyclone, which generates some of the worst tornadoes.

The death of Finley's tornado-forecasting program left Americans almost helpless before the tornado threat. That helplessness was epitomized by the worst tornado in American history, the so-called Tri-State Tornado in Missouri, Illinois, and Indiana on March 18, 1925. The disaster actually consisted of several tornadoes that traveled a total of 219 miles, killing 689 people (including 234 in Murphysboro, Illinois) and injuring more than 2,000. Americans are still alive today who recall the tempest's viciousness. Writing in a 1995 issue of *American Heritage*, Howard E. Rawlinson of Mt. Vernon, Illinois, says he was in high school when the school janitor announced, "Boys, if you've never seen a tornado, you're going to see one now." Two of Rawlinson's cousins were tending potatoes in the cellar of their home, and one walked upstairs at the moment the tornado hit. It slammed the house to the side, decapitating him. "The younger brother huddled in the darkness of the cellar, a white leghorn chicken roosting

High-speed upper air winds

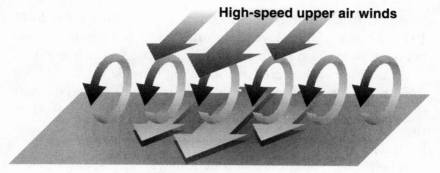

Slower low level winds

Warm updraft

Joe Shoulak

THE FORMATION OF A MESOCYCLONE

on his shoulder." Neighbors searched for a missing woman; they found her naked body in a tree. "The most gruesome sight I recall was the body of a Jersey cow that had been run through by a two-by-four . . . an occasional oak or elm or sycamore held out a bone-white stump of an amputated limb. . . ."

Rural folk were deeply religious. Why, they wailed, would

God permit such ghastliness? After the St. Louis tornado of 1927 (72 dead, 500 injured), an anxious local preacher urged his flock to look on the bright side, to perceive the tragedy as "a visitation from a merciful and loving Providence. . . . Whom the Lord loveth he chastiseth. Chastisement here is better than chastisement hereafter." His view was cold comfort to the victims, some of whom would suffer from their injuries for the rest of their days.

The 1930s brought the Depression, war scares, and terrible weather to America. One of the decade's ghastliest tornado outbreaks struck the South on the evening of April 5, 1936. Elvis Presley was fifteen months old when an incredible tornado ripped through his hometown of Tupelo, Mississippi. His mother, Gladys, "clung to her baby and huddled in fear in their small house," says biographer Patricia Jobe Pierce. "The tornado leveled St. Mark's Methodist Church across from the Presleys' home and flattened other shacks along the street . . . years later Gladys convinced Elvis that God saved him that day (and the day he was born) because God had determined Elvis was 'born to be a great man.' "

The F5 twister killed at least 216 people and injured 700. It was part of a major tornado outbreak across the South. The Tupelo twister was "one of the deadliest in U.S. history," Grazulis says. "Entire families were killed, up to 13 in a single home. . . . A movie theater was turned into a hospital with the popcorn machine used to sterilize instruments." The true death rate may have exceeded 216: The tornado injured many blacks, but in that place and time, "only the names of the white injured were published in newspapers."

The Tupelo outbreak was the climax of the "dark age" of American tornado science. It was the last catastrophic American tornado outbreak before weather forecasting was transformed by two global tumults: World War II and the early Cold War.

Tragically, important scientific advances often stem from military research. During World War II, accurate weather

TWISTER

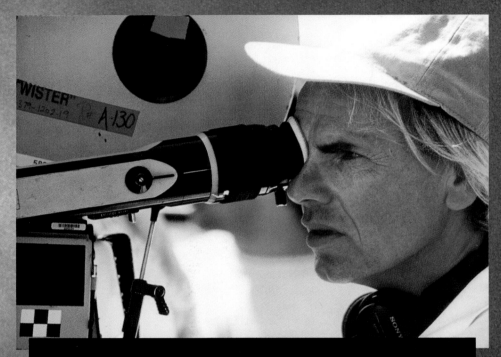

Director Jan De Bont composes his shot at the first tornado's irrigation-tion ditch encounter. [DAVID JAMES]

Actors and camera crew are pummeled by the special effects crew's hail and debris for the "Hailstone Hill" sequence. [DAVID JAMES]

The small town of Wakita, Oklahoma, is "dressed" for the night sequence approaching Aunt Meg's house. [DAVID JAMES]

A 3,500-pound replica of a real tank truck being raised and prepped for the big explosion. [RON BATZDORFF]

Special effects crews use gasoline and primer cord to get the explosion of the tank truck at the F5 tornado. [RON BATZDORFF]

Jo and Bill Harding (Helen Hunt and Bill Paxton) watch as the F5 tornado grabs the dorothy and sucks up the sensors. [DAVID JAMES]

Jonas's team preps him for his tornadic encounter. Left to right: Patty (Melanie Hoops), Dean (Dean Lindsay), Tony (Anthony Rapp), Jake (Jake Busey). [DAVID JAMES]

In southwestern Oklahoma on the last week of VORTEX '95, Jerry Straka gazes at gathering storms from his and Josh Wurman's "Doppler on Wheels" mobile radar vehicle. [JIM REED]

A twister-toppled tractor-trailer lies on its side like a dead dinosaur in Corum, Oklahoma, on April 17, 1995. The driver was unhurt. [JIM REED]

A huge supercell thunderstorm near Laverne, Oklahoma, spawns a tornado just after sunset in May 1991. The twister passed through the Laverne outskirts and caused minor structural damage. [WARREN FAIDLEY]

A tornado passes through open country five miles south of Clear-water, Kansas, on May 16, 1991. [KEITH BREWSTER]

Multiple lightning bolts strike a southeastern Arizona city. [WARREN FAIDLEY]

forecasts were required to plan troop movements and to time bombing raids. So the military spent fortunes developing ways to gather more weather data. Meteorology also benefited from military actions that originally had nothing to do with weather. The development of rocket science would lead, after the war, to space satellites, including weather satellites; and the invention of electronic computers (for plotting artillery trajectories and designing the first atomic bombs) would lead to computerized weather forecasts. For meteorologists, the war's biggest boon was radar. Radar, for radio detection and ranging, was meteorology's first major "remote sensing" tool.

Bomber pilots crossing the Atlantic made the war's top atmospheric discovery: the "jet stream," a high-altitude, ever-meandering river of extremely strong winds (hundreds of miles per hour). The jet stream helps to energize and steer surface storms. Meteorologists increasingly used "radiosonde" balloons—battery-powered balloons with weather instruments and radio transmitters—to measure wind speeds and directions tens of thousands of feet above Earth. Understanding the current position and intensity of the jet stream and other high-altitude winds is vital for tornado forecasting.

During the war, the scientists Albert K. Showalter and Joe R. Fulks identified the weather factors that tend to precede thunderstorms. These include (in paraphrase):

—A rapid temperature decrease with height. This encourages convection and, in turn, thundercloud formation.

—A distinct boundary between high surface humidity and a large, dry region high above. This spurs moist surface air to convect into the upper air, where it forms clouds.

—A "temperature inversion," or warm layer of air over cooler surface air. Normally surface air is warmer than the upper air, but a temperature inversion reverses the situation. The inversion blocks the rise of warm, moist air (like a lid on a boiling pot). If the inversion is breached, the long-confined warm air may gush through the opening into the upper air and rapidly initiate thunderstorms. The breach may be triggered by a passing cold front, which shoves warm air through the inversion.

World War II ended in 1945. A year or two later, the Cold War began. The military kept some of its long-range missiles and bombers in the Midwest, where they were less vulnerable to Soviet attack than on either coast. But Midwestern bases faced a very different threat, as the generals learned on March 20, 1948.

Early that day, at Tinker Air Force Base near Oklahoma City, the base's stocky, bespectacled meteorologists—Major E. J. Fawbush and Captain Robert C. Miller—forecast thunderstorms. At 10 P.M., a half-mile-wide tornado hit Tinker. While the tornado twisted and overturned millions of dollars worth of aircraft, Fawbush and Miller huddled in the base storm shelter and lamented their shortcomings.

Fawbush had been embarrassed before. A decade earlier he had worked at a Louisiana military base. One day he forecast "nice, mild weather." That night as he slept, a twister passed two blocks away. "That sort of got my goat," he later admitted. Now Mother Nature had gotten his goat again, and much more dramatically: The Tinker tornado wrecked 52 big aircraft, including 17 C-54 transports, two B-29 bombers, and 15 P-47 fighters. Stalin himself might as well have bombed Tinker.

Base officials ordered a board of inquiry. Such a catastrophe must never happen again. But how could it be prevented? Fawbush and other weather experts reminded the board that tornadoes were highly unpredictable. A meteorologist might forecast severe storms for a wide region, but he couldn't anticipate the exact target of a twister—say, a specific military base.

Five days after the Tinker tornado, Fawbush and Miller reviewed U.S. weather developments and realized that another storm was headed their way. It seemed highly unlikely that a second tornado would hit the base. Still, they didn't want to risk another flubbed forecast, so they relayed their concerns to base officials. Crews tied down all vulnerable airplanes or moved them into hangars.

Incredibly, at 6:00 P.M. that day *another* tornado hit Tinker. The funnel, a "white finger," dropped from a greenish-black cloud, flattened hangars, and chewed up 35 more planes. Five days earlier, Fawbush's and Miller's names had been mud; now they were heroes. Their lucky forecast marks the dawn of modern tornado science.

Afterward, Fawbush and Miller toiled (sometimes all night) to refine their forecast technique. Like Showalter and Fulks, they believed that severe weather was preceded by large-scale (synoptic) weather trends. To prove their case, they pored through decades' worth of weather records stored on microfilm and punch-card files (an early form of computerized data storage). Their goal: to see if specific synoptic conditions preceded not just thunderstorms but *tornadic* thunderstorms. Their conclusion: yes. Most tornadoes were preceded by a tongue of warm, moist surface air; dry western winds at a higher altitude; a powerful jet stream blowing at an angle to surface winds; and a front that forced the warm air upward.

In the summer of 1949, they successfully forecast 18 Midwestern tornadoes. Whenever they expected a severe storm, they used a teletype machine to transmit a coded warning to Air Force bases. The code word for hail was "orange"; for strong winds, "yellow," and for tornado, "black." The next January, they presented their findings at the American Meteorological Society meeting in St. Louis.

Their achievement sparked a minor sensation, especially in Tornado Alley. The *Saturday Evening Post* praised Fawbush's and Miller's work, noting that farmers "will now have ample time to walk—not run" to their cellars after hearing a tornado warning. Had humans finally found a way to anticipate the twister's wrath?

In the mid-1930s, English newspapers reported peculiar news from the countryside. Someone was building odd-looking buildings with mysterious antennae. Writers proposed that scientists were developing "death rays."

In reality, the buildings were secret installations for testing a new weapon. Radar (for "radio detection and ranging") detected distant airplanes by emitting microwave radiation, which bounced off the planes and returned to the radar's receiving antenna. Radar later helped the Allies to defeat Hitler's bombers. "The [atomic] bomb ended the war," said one top U.S. scientist, "but radar won the war."

Early in the war, radar operators noticed that radar also detected rain and hail. Meteorologists got excited: Maybe they could use radar to spot distant storms! After the war, the Weather Bureau and other research institutions bought military-surplus radar units.

About 5:15 P.M. on April 9, 1953, a severe tornado passed near Urbana, Illinois. The F3 twister traveled 38 miles and carved out a 150-yard path. It killed 1 person, injured 10, wrecked 8 homes, and damaged 72. By luck, a radar unit was operating at the Illinois State Water Survey. No one watched the screen during the storm, but an automatic 35 mm camera recorded the radar images. Afterward, analysts realized the images showed a peculiar echo: It was shaped like a hook or the number 6. The hook followed the same path as the tornado. Famed tornado researcher Tetsuya Theodore Fujita of the University of Chicago analyzed the radar pictures and concluded the hook wasn't the tornado funnel itself. Rather, the hook represented a much wider, vertical column of rotating air about 30 miles wide. The rotating column resembled "a miniature hurricane in many respects," Fujita said. (The "miniature hurricane" was what we today call the mesocyclone.)

The Urbana achievement couldn't have come at a more disastrous time. The year 1953 brought some of the worst tornadoes in American history. On May 11, a tornado slammed downtown Waco, Texas, killing 114 people, injuring 597, and destroying 200 businesses and 150 homes. The city's downtown was a mountain of rubble. A six-story furniture store collapsed. "I saw roofs pop up like corks from champagne bottles," a witness later wrote. "I saw automobiles—some unoccupied, others bearing passengers—crushed like bugs and buried out of sight under great mounds of brick, timber, steel, and glass. I saw one car, an old model, leap upward and disappear into the air as if by magic."

The Waco catastrophe marked "the first time that a major tornado struck the business section of a large city (population 90,000) during business hours," Edward Brooks wrote in *Weatherwise*.

Soon two more fair-sized cities were hit—both in states where severe tornadoes are rare. On June 8, a tornado killed 115 people in Flint, Michigan. On June 9, a "huge cone of black smoke" hit Worcester, Massachusetts; it killed 94 and injured 1,288. The sight "was incredible to these New England eyes," a Worcester newspaperman said. Two Worcester mothers reportedly pulled down a child who was floating away. One witness "awestruck" by the damage was an eleven-year-old boy, Thomas P. Grazulis, who later became the nation's leading tornado climatologist. Thirty-five miles east of Worcester, odd objects fell from the sky: a French music box, a large aluminum trapdoor, and a couch cover that was "frozen solid."

Americans had plenty of fears in those days—fear of Soviet attack, fear of communist spies, fear of hydrogen bombs. And now, it appeared, they had a new fear: tornadoes. Tornadoes were once a mainly rural phenomenon; now they were headed for Main Street. Moviegoers gasped at black-and-white newsreels of the destruction of Waco, Flint, and Worcester, and wondered if their town was next.

The night of the Worcester disaster, Representative Ray J. Madden of Indiana went on national radio and made a startling request: The Senate and House Armed Services Committees should investigate whether atomic bomb tests had triggered the tornadoes. "The number of these death-dealing twisters this year is unprecedented in Weather Bureau annals.

"The public is entitled to know," he declared, "if there is

any connection between the A-bomb explosions and the tornadoes which have followed. Scientists admit that great quantities of radioactive materials are blown into the stratosphere by the explosions of atomic bombs in Nevada. Thousands of tons of this radioactive material drift across the nation. We know that this material pollutes the air we breathe and the water we drink. Does it also generate abnormal atmospheric conditions?"

The United States had been testing atomic bombs in Nevada for more than two years. Colonel Ben Holzman of the U.S. Air Force denied the bombs caused the tornadoes. A "very small-sized but moderate warm front rainstorm releases energy at the rate of at least three atomic bombs per second," he said. Another skeptic was Albert Showalter, who ran the Weather Bureau's Analysis Center: "Almost every bit of unusual weather is systematically connected with any unusual event." D. Lee Harris of the Weather Bureau conducted a study and announced there was little correlation between rainfall and paths of radioactive clouds.

Nonetheless, the media ran with the story. *U.S. News & World Report* ran a cover story, "What's Wrong with the Weather," which featured an unusually long, ten-page interview with U.S. Weather Bureau chief Francis W. Reichelderfer. "What does science really know about the atom bomb's effect on weather?" the magazine asked. "How are destructive tornadoes to be explained?" The balding, stark-eyed Reichelderfer was reassuring. While it was true that the atomic blasts ejected dust into the atmosphere, that dust wouldn't have caused rainfall across the country. He noted that vast dust clouds from the "Dust Bowl" of the 1930s reached Washington, D.C., in amounts a billion times as great as particles from the atomic tests. Yet the Dust Bowl clouds were never associated with excess rain.

Behind the scenes, the Weather Bureau took political advantage of the public's tornado jitters. Bureau officials asked Congress for a big boost in funds to improve tornado and hurricane warnings. Rattled by photos of rubble-strewn towns,

Congress okayed the increase. One of the Bureau's first acts was to install weather radar units across Tornado Alley and along the hurricane-prone coast.

In the fight against twisters, radar scored its first big success in April 1956. At Texas A&M in College Station, Texas, researchers saw an approaching thunderstorm on a local radar unit. Analysts estimated that the storm would hit town about the same time that students left public schools. They contacted school officials and urged them to keep students in classes until the storm passed. The storm dropped a twister one block from a school. If students had been walking home as the funnel dropped, it could have slaughtered them.

We have come to realize that these vicious storms [tornadoes] are much more frequent and more widespread than used to be thought. They are most common in the Midwest, but tornadoes are now known to strike in every part of the country and at every season.
—Morris Tepper, _Scientific American_ (1958)

*F*rancis Reichelderfer was a visionary. Then in his late fifties, a former naval officer who oversaw the 17,000 employees of the Weather Bureau, he championed the early use of electronic computers for forecasting. Later he promoted the development of weather satellites. His main contribution to tornado science came in 1952, when he opened the Weather Bureau's Severe Local Storm Warning Center (SELS) in Washington, D.C.

SELS was "a tense, young agency set up . . . to warn any part of the nation when twisters are on the prowl," the _Saturday Evening Post_ observed. SELS was also controversial. In the early 1950s, forecasters were still reluctant to use the word "tornado." "They were afraid we would scare somebody. The surest way to get fired was to use the word 'tornado' in a forecast," one forecaster told a reporter. Reichelderfer hired Kansas City forecaster Donald C. House to run SELS. House accepted the job reluctantly: "I was afraid that I was in for a creepy job." But 1953's tornadoes—Waco, Flint, Worcester—convinced House that SELS could make a difference. He persuaded Reichelderfer to transfer SELS to Kansas City, the heart of Tornado Alley.

At 11:00 P.M. Eastern time on March 17, 1952, SELS issued its first tornado "watch": "There is a possibility of tornadoes in eastern Texas and extreme southeastern Oklahoma tonight, spreading into southern Arkansas and Louisiana before daybreak."

The watch stayed in effect until 7:00 A.M. No one saw any tornadoes.

SELS issued its first successful tornado forecast four days later. A major tornado outbreak struck Arkansas, Tennessee, Alabama, Missouri, Kentucky, Louisiana, and Mississippi. Twisters killed 208 people and injured more than 1,000.*

To improve tornado warnings, SELS acquired a "Tornado Research Airplane." In 1956 SELS contracted with a pilot, Jim Cook of Jacksboro, Texas, to fly his single-engine F-51 aircraft into stormy regions. Cook was thirty-four, a towering, handsome "Right Stuff" type who spoke with a drawl and a certain macho nonchalance. The press treated him as a hero—imagine, someone who flew *toward* tornadoes! "The first man ever deliberately to court tornadoes," *Newsweek* called him. Indeed, he was almost certainly the first serious "storm chaser." (Famed storm chaser David Hoadley didn't start chasing storms—by car—until 1957.) Cook was tough, but no fool. He avoided greenish thunderclouds because he

---

* SELS was the ancestor of today's National Severe Storms Forecast Center in Kansas City. NSSFC is the nation's first line of defense against tornadoes and other local severe storms. NSSFC meteorologists monitor the nation's weather 24 hours a day, 365 days a year, watching for large-scale changes—say, the birth of a mean-looking cold front, or an ominous shift in the jet stream—that could spawn thunderstorms, blizzards, floods, or other violent weather. (Hurricanes are handled by the National Hurricane Center in Miami.)

believed they contained hail: "If you play too close," he drawled, "sooner or later you'll plow up a snake. That's not for me." He made his first tornado flights out of Dallas in April 1956. His plane carried instruments including thermometers, humidity gauges, air pressure instruments, wind gauges, and 35 mm tracking cameras. He recorded his weather observations on a dictation machine.

Cook's flights saved lives. For example, on a 1961 flight in a war surplus P-38, he spotted a funnel cloud over Oklahoma. It dangled (in the *Post*'s words) "like a dirty, oily rope." The funnel raced toward a town, "one of those quiet prairie places with a main street, a river, a railroad track, and a grain elevator. He was low enough to see children playing and people walking about. Obviously no one down there knew of the tornado. . . ." Cook radioed the Tulsa airport, hoping officials there could warn the town. Unfortunately, he didn't know the town's name. He struggled to explain to an air traffic controller that "it's near Henryetta and it lies near two little lakes." But electrical static drowned out his transmissions.

Time was running out, so he elected to warn the town himself. He lowered his P-38 and flew it directly down the town's main street at 200 miles per hour. The racket startled locals, who looked up and pointed. Then he veered the P-38 toward the tornado. As townspeople viewed his ascent, they noticed the funnel. He watched with relief as they rushed to backyard cellars and slammed the doors shut.

For the first time since Sergeant John Finley's brief, failed effort, the U.S. government was at war against tornadoes. Humans, the pursued, were now the pursuers, who chased tornadoes to dissect their secrets. But a handful of visionaries wanted to move beyond dissection, to the next logical step: extermination.

CHAPTER 4

"To Destroy Tornadoes Before They Destroy Us"

By the 1950s, Americans were no longer utterly helpless before twisters. To anticipate tornado-breeding storms, meteorologists tracked large-scale movements of air masses, temperature inversions, fronts, and other weather conditions. Radar networks rose across the countryside, providing early warning of severe storms. Aircraft prowled the Midwestern skies, looking for tornadoes. An exotic new warning system would emerge at the start of the next decade, when the first weather satellites soared into orbit. Their black-and-white photos allowed meteorologists to study the evolution of thunderstorms, some of which spawned twisters. In short, Sergeant John Finley's nineteenth-century dream was coming true, thanks to technologies undreamed of in his day.

But a key unknown loomed: the nature of the tornado funnel itself. How did so small an object—rarely more than a mile wide—acquire such ferocious energy? Could "ordinary" physics explain its violence? Or was some more exotic explanation required, one that transcended classical atmospheric science?

Also fascinating was the question of the tornado's internal structure. How exactly did winds flow within the funnel, which a whirling shroud of dust and debris concealed from human gaze? These questions were driven by practical concerns—for example, what would happen if a twister hit a nuclear power plant? Is there something about tornadoes' internal dynamics that causes them to avoid large cities? And could we develop a weapon to *stop* tornadoes, a weapon that would (in one questioner's words) "destroy [them] before they destroy us"?

The disastrous urban tornadoes of 1953 convinced some observers that the United States should no more let tornadoes run rampant over the nation than let Soviet tanks roll over Europe. That year, on November 12, two startling speeches were given at an American Meteorological Society (AMS) meeting at the University of Texas in Austin. The speakers, Colonel Rollin H. Mayer of the U.S. Air Force and physicist Fritz O. Rossmann, called for an all-out war against twisters, using two relatively new technologies: radar and the guided missile.

"Statistics show that within the past 20 years approximately one billion dollars property damage, more than 8,000 deaths, and over 100,000 persons have been injured as a result of tornadoes," said Mayer, an electronics expert at the Air Force Missile Test Center at Patrick AFB on Cape Canaveral, in a published summary of his remarks. "If only a 5 percent reduction in tornado damage could be realized, a great advancement in science will have been made and the welfare of mankind promoted."

The previous April, the Urbana, Illinois, radar had detected the "hook" echo from a tornadic storm. Mayer suggested building a national network of weather radar. It would detect tornadoes, just as the Distant Early Warning (DEW) radar network—then being developed on the fringes of the Arctic Circle—would spot incoming Soviet bombers.

If the network spotted a tornado, he continued, then the Air Force could launch a guided missile at it. Radar would guide the missile to its target. The missile's explosive warhead (possibly an atomic bomb) would detonate the funnel. "A practicable tornado destructor might be realized in the near future," Mayer claimed.

One can imagine how the assembled meteorologists reacted to Mayer and Rossmann's scheme. Thirty years before President Reagan called for building the "Star Wars" anti-ICBM defense system, they were calling for something similar, except it would hit tornadoes. Of course, an anti-tornado missile would need to be steered with exquisite precision, lest it veer off course and hit the wrong target—such as the town it was intended to protect. Their strategy was within the technological capability of the day, however. The Pentagon was already developing radar-guided missiles, such as the Nike, that would shoot down Soviet bombers. It was also developing small ("suitcase") nuclear weapons that could generate blasts weaker than those that leveled two Japanese cities.

In the meantime, Mayer said the nation could develop "a fleet of airplanes loaded with missiles waiting to attack tornadoes. . . . We may not be able to knock out the big tornadoes right away, but in the near future we may be able to handle the little ones." The tornado warning network would consist of "mobile anti-tornado stations, assembled at airports near large cities where tornadoes might be expected. . . . Into the net would be fed tornado forecasts, radar detection, warning, and tracking data. Jet planes with tornado-destruction missiles would be standing by to destroy tornadoes before they destroy us."

Weather experts reacted ambivalently to Mayer and Rossmann's proposal. On the one hand, "the Weather Bureau has an open mind on this," said a Bureau staffer in Washington. On the other hand, meteorologists questioned the feasibility of knocking down tornadoes as if they were so many Russian bombers. Little was known about tornadoes' cause and behavior. Some were true aerial monsters, thousands of feet wide, with winds that some experts thought were supersonic. Even an atomic bomb might be helpless against such titans.

The anti-tornado defense system was only one of the spectacular "weather modification" schemes considered in the 1950s. Respected researchers talked about transforming deserts into gardens, diverting hurricanes from U.S. coastlines, and spawning thunderstorms to wash pollution from Los Angeles skies. "Weather control" was an old dream, of course. The United States had a long and sordid history of such dreams; nineteenth-century hucksters gulled farmers with costly and useless schemes to end drought and prevent hail. But weather control turned scientific in the late 1940s, thanks to three scientists at General Electric—Vincent J. Schaefer, Irving Langmuir, and Bernard Vonnegut. Schaefer found a way to unleash rainfall by dropping small amounts of dry ice into clouds. The dry ice (frozen carbon dioxide) initiated condensation of water droplets into larger droplets, which were heavy enough to fall to Earth. Later Vonnegut realized one could get the same effect with silver iodide, which, unlike dry ice, doesn't require refrigeration.

"It becomes apparent that important changes in the whole weather map can be brought about" with cloud seeding, said Langmuir, a Nobel laureate who ran the General Electric laboratory. He claimed that cloud seeders had already inadvertently caused a hurricane to alter its path. "With increased knowledge, I think we should be able to abolish the evil effects of these hurricanes." He also maintained that seeding experiments in the Southwestern U.S. had upset weather conditions on the East Coast.

Weather Bureau experts disagreed. They conducted statistical analyses of cloud seeding experiments and concluded they had little effect on rainfall. The hurricane probably changed course by coincidence. Their doubts riled Langmuir, who "was a very emotional, very touchy guy," recalls National Oceanic and Atmospheric Administration's (NOAA's) Joanne Simpson, a noted weather modification researcher. "I went to a conference in the 1950s and he and another scientist were screaming at each other. . . . I liked him very much, but he got very paranoid when the Weather Bureau went after him to demand 'evidence.' "

Popular interest in weather modification "rivaled that engendered by the announcement of the discovery of nuclear fission," thunderstorm researcher Horace R. Byers later said. U.S. Senator Clinton P. Anderson proposed a "weather control bill." Should weather-modification technology fall into the wrong hands, he said, the implications were "almost terrifying." General George C. Kenney, former head of the Strategic Air Command, warned: "The nation which first learns to plot the paths of air masses accurately and learns to control the time and place of precipitation will dominate the globe."

In this overwrought atmosphere, it's no wonder that Mayer and Rossmann weren't laughed out of the American Meteorological Society meeting. Just a month after their speeches, Captain Howard T. Orville (U.S. Navy, retired), chair of President Eisenhower's advisory committee on weather control, said it was too early to say whether cloud seeding would lead to tornado or hailstorm control—but it was possible.

But was it desirable? On the one hand, the possibility of breaking up a tornado "seems quite remote," wrote Morris Tepper, a leading tornado expert, in *Scientific American* in 1958. "Even if we could predict exactly where tornadoes may form, nothing short of a massively energetic intervention (such as huge thermonuclear explosions) could influence the great volume of air involved, which amounts to thousands of cubic miles. And of course such a cure would be worse than the disease—to say nothing of the possibility that our interference might trigger off tornadoes that would not otherwise have started."

On the other hand, Tepper speculated, cloud seeding might "soften the fury" of a tornado. Later, other researchers investigated the possibility of using cloud seeding to weaken updrafts that feed tornadoes, or to trigger cold downdrafts that would extinguish them as a stiff breeze extinguishes a candle.

Nothing ever came of Mayer and Rossmann's colorful scheme. It was quietly forgotten, along with other fantastic technological visions of the Eisenhower era such as nuclear-powered airplanes and the colonization of the ocean floor. But other tornado-control notions would soon emerge. The most exotic involved an everyday phenomenon: electricity.

How a tornado develops its prodigious energy is still a complete mystery.
—Morris Tepper, *Scientific American*
(August 1958)

A major problem in explaining the mechanics of tornadoes is to find the source of the kinetic energy and to account for its concentration over a rather small space.
—A. J. Abdullah, *Monthly Weather Review*
(1955)

The tornado may originate as a small ordinary whirlwind, later becoming intensified by the electrical energy released by a severe thunderstorm. . . . Once the tornado and its mechanism is understood, man will be better able to predict its impending birth, and perhaps, ultimately, even prevent its formation.
—Peter E. Viemeister, *The Lightning Book*
(MIT Press, 1961)

We may someday learn to turn tornadoes off before they begin by some process of de-electrifying the clouds which are likely tornado producers.
—Robert G. Watts Jr., Tulane University engineer
(1972)

"**B**aby boomers" grew up watching television reruns of the 1939 movie *The Wizard of Oz*. No matter how many times this writer saw the film, he gasped every time he saw Dorothy and Toto's home sucked into the tornado funnel. It was an impressive special effect by the standards of the 1930s, and it is still convincing today.

In real life, many people have ventured into tornadoes. The best-known tale is told by Will Keller, a Kansas farmer. On

June 22, 1928, a tornado hit his home. He smelled a "strong gassy odor," then

> I looked up, and to my astonishment I saw right into the heart of the tornado. There was a circular opening in the center of the funnel, about fifty to one hundred feet in diameter and extending straight upward for a distance of at least half a mile, as best I could judge under the circumstances. The walls of this opening were rotating clouds and the whole was brilliantly lighted with constant flashes of lightning which zigzagged from side to side.

On May 3, 1948, a tornado struck McKinney, Texas, killing 3 people and injuring 43. The survivors included Roy S. Hall, a retired U.S. Army captain, and his wife and baby. Writing in *Weatherwise* magazine, he described how the tornado blew him and his family out of their home. Sprawled on the ground, he looked up and saw "the lower end of the tornado funnel":

> The bottom of the rim was about 20 feet off the ground, and had doubtless a few moments before destroyed our house as it passed. The interior of the funnel was hollow; the rim itself appearing to be not over 10 feet in thickness and, owing possibly to the light within the funnel, appeared perfectly opaque. Its inside was so slick that it resembled the interior of a glazed standpipe. . . . The whole thing was rotating, shooting past from right to left with incredible velocity.
>
> I lay back on my left elbow, to afford the baby better protection, and looked up. It is possible that in that upward glance my stricken eyes beheld something few have ever seen before and lived to tell about. I was looking far up the interior of a great tornado funnel! It extended upward for over a thousand feet, and was swaying gently, and bending slowly toward the southeast. Down at the bottom, judging from the circle in front of me, the funnel was about 150 yards across. Higher up it was larger, and seemed to be partly filled with a bright cloud, which shimmered like a fluorescent light. This

brilliant cloud was in the middle of the funnel, not touching the sides. . . .

Keller's and Hall's tales intrigued Bernard Vonnegut, who had co-pioneered cloud seeding with Schaefer and Langmuir. An expert on atmospheric electricity, Vonnegut was especially interested in Keller's and Hall's description of what might be unusual electrical events within the tornadoes—the "constant flashes of lightning" and the "brilliant cloud." Vonnegut himself had witnessed the thunderstorm that spawned the murderous Worcester, Massachusetts, tornado of 1953. They generated "the most spectacular lightning I had ever seen . . . at least 20 lightning flashes per second." He calculated that the storm had generated about 100 million kilowatts of electricity, "roughly equivalent to the generating capacity of the entire United States." Could lightning be the covert fuel source for tornadoes?

It was an old idea. The science of atmospheric electricity was born in 1752, when Benjamin Franklin flew history's most famous kite into a thunderstorm. The kite hung from a silk thread, at the end of which dangled a metal key. During the storm, Franklin reached for the key. An electric spark leaped between the key and his hand. That spark proved that lightning is ordinary electricity. Anyone can generate electricity by walking across a rug, then reaching for a doorknob: A electric spark leaps between your hand and the knob.

Franklin's work inspired the sale of lightning rods to protect buildings. Some regarded the rods as sacreligious. One science lecturer of the day felt compelled to reassure his audience that "the erection of lightning rods is not . . . inconsistent with any of the principles either of natural or revealed religion."

Because lightning is so brilliant and explosive, some early scientists thought it triggered another explosive phenomenon: tornadoes. "After maturely considering all the facts, I am led to suggest that a tornado is the effect of an electrified current of air superseding the more usual means of discharge between the earth and clouds, in those sparks or flashes which are called lightning," the noted chemist Robert Hare wrote in 1837.

Others were skeptical. Hans Christian Oersted, a Danish physicist who pioneered the study of electromagnetism, asked: If tornadoes were electrical, then why didn't sailors suffer electrical shocks when they passed near waterspouts? And why didn't the waterspouts violently deflect the ships' magnetic needles, as they should if electricity surged through the funnels?

The debate ended up in court. In 1887, tornado expert John P. Finley of the Signal Service testified in an insurance case. A tornado had struck the insurant's home. The insurance policy only covered lightning damage. However, the insurant claimed that lightning spawned tornadoes, therefore the insurance company should pay up! In his testimony, Finley cited more than 140 objections to the electrical theory of tornadoes. The insurance company won the case.

Afterward, the electrical theory of tornadoes was largely forgotten for almost seventy years—until it was rediscovered in the 1950s by Bernard Vonnegut. Through his long and sometimes controversial career, Vonnegut has pursued many offbeat ideas. Some are so offbeat that they'd seem at home in a novel by his younger brother, the satirist and science-fiction writer Kurt Vonnegut Jr. For example, Bernard was fascinated by farmers' reports that tornadoes defeathered their chickens. Could one estimate a tornado's wind speed by how thoroughly it defeathered a chicken? To find out, he placed chickens in a wind tunnel and turned on the fan. The fan blew off the feathers, but in an inconsistent manner. He concluded that defeathering was no substitute for a sturdy anemometer. He published his findings in a short scientific paper, entitled "Chicken Plucking as a Measure of Tornado Wind Speed."

In Kurt's autobiographical prologue to his novel *Slapstick* (1976), he affectionately depicts Bernard's laboratory as "a sensational mess . . . where a clumsy stranger could die in a thousand different ways, depending on where he stumbled." A safety officer once complained about it to Bernard, who good-humoredly tapped his forehead and replied: "If you think this laboratory is bad, you should see what it's like in *here.*" Kurt recalled Bernard showing him a device that "clicked" as it detected electrical activity in a distant cloud. "I thought it was beautiful that my big brother could detect secrets so simply from so far away."

The Vonnegut boys grew up in Indianapolis in a respected German-American family of some means—means that, come the Depression, largely disappeared. Despite the family's financial struggle, Bernard won admittance to MIT, where he received his doctorate in physical chemistry at age twenty-five. In the 1940s he became an atmospheric physicist at General Electric Corporation in Schenectady, New York (where Kurt worked in public relations). There, along with Schaefer and Langmuir, Bernard co-pioneered cloud seeding. Since 1967 he has worked at the State University of New York in Albany. Now in his early eighties and semiretired, he says, "I'm skeptical about a *lot* of things, and I think that's a good way to be."

He has long doubted the traditional theory of how tornadoes form (tornadogenesis). The atmosphere is a fluid, like water. In theory, atmospheric behavior can be explained with equations from the science of fluid dynamics, as can the behavior of any other fluid—say, water going down a bathtub drain. For decades scientists have tried to figure out how the fluid motion of the atmosphere, fueled by heat and rising water vapor, generates the ghastly violence of tornadoes. They haven't fully succeeded, despite much progress.

In 1960, Vonnegut suggested they were on the wrong track. In "Electrical Theory of Tornadoes," an article for the *Journal of Geophysical Research,* he calculated that the tornado updraft is much too strong to be blamed on ordinary temperature differences in the air. Something else must be driving the updraft. That "something" could be lightning or other manifestations of atmospheric electricity. Perhaps lightning bolts repeatedly flashed in the same place, heating it and creating intense convection—that is, an updraft—in a narrow column. Start the column rotating, and you have a tornado. (Alternately, he said, the storm's electric field might accelerate electrically charged atoms to high speed.)

Skeptics scoffed. Ross Gunn of the U.S. Weather Bureau pointed out that if Vonnegut were right, then lightning should blaze around tornadoes. But in Kansas, electrical field-measuring stations run by "public-spirited farmers" had detected little unusual electrical activity near funnels. In 1961, F. C. Bates and his assistants from the University of Kansas studied a thunderstorm from an aircraft and observed that "at no time in the [tornado] life cycle did intense lightning discharges take place beneath the parent cloud."

Undiscouraged, Vonnegut continued to promote his lightning theory. He publicized a puzzling photo taken in Toledo, Ohio, on April 11, 1965. That night, a tornado hit the city, killing 16 people. Unaware of the twister, James Weyer photographed what appeared to be two vertical shafts of light passing over Toledo. Vonnegut published the photo in the journal *Science*. He suggested the columns were "tornado funnels illuminated by some type of electrical discharge." His claim triggered a minor scientific brouhaha. Critics proposed many alternative explanations. Some suggested the columns were ordinary funnel clouds that reflected city lights. Others proposed the "shafts" were chemical flaws on Weyer's film, or light reflections within his camera. (The latter often cause "UFO" images in photos.)

Vonnegut countered by heading for his laboratory. He and a colleague heated a column of air by generating high-voltage discharges. The result: "a miniature tornado-like vortex." He concluded "the heating produced by electrical discharges in a large storm may play a significant role in forming and maintaining natural tornadoes."

Where was all this leading? In his 1960 paper, Vonnegut offered a prophecy:

> If tornadoes are caused by electrical mechanisms, then
> through better understanding of these mechanisms it should
> be possible to increase the accuracy of tornado forecasts, to
> obtain warning of their presence, and, conceivably, to inhibit
> or prevent their formation.

Vonnegut never tried to "inhibit or prevent" tornadoes
himself. But his idea inspired two scientists to consider doing
so: Vernon J. Rossow and Stirling Colgate.

The problem was this: If lightning was the secret behind
tornadoes, then one might suppress tornadoes by suppressing
lightning. By coincidence, other scientists were *already* trying
to suppress lightning—not because they wanted to stop tornadoes, but because lightning itself is a major natural hazard.
A lightning-suppression program called "Project Skyfire" had
been operated since the early 1950s by several federal agencies, including the Weather Bureau, the Departments of Interior and Agriculture, the National Science Foundation, and
the President's Advisory Committee on Weather Control. They
hoped to prevent lightning from starting forest fires and hitting sensitive facilities such as airports, missile bases, and nuclear weapons storage sites.

Lightning occurs when different parts of a thundercloud
develop opposite electrical charges (negative and positive).
Then a huge spark (a lightning bolt) leaps between the negative and positive areas, neutralizing them. Scientists reasoned
that if they could *slowly* neutralize the cloud, then the charge
differences might remain too slight to trigger lightning. (Likewise, one might prevent a dam break by slowly draining water
from the dam.) So the lightning suppressors seeded thunderclouds with "chaff"—tiny aluminum strips that might slowly
discharge the storm's electricity. They also launched small
rockets that towed long wires into the storm. In theory, the
wires could link regions of opposite electrical charges,
thereby closing the electrical circuit, neutralizing the thundercloud and preventing lightning. Such techniques were
tested during storms, with ambiguous results.

The September 22, 1967, issue of *Science* ran a paper titled, "Tornadoes: Mechanism and Control." Its author was a forty-two-year-old physicist, Stirling A. Colgate, a veteran of the U.S. hydrogen bomb and nuclear fusion energy programs. Colgate's paper began: "The possibility that the energy source of tornadoes is primarily electrical has been suggested independently by several authors, the most convincing analysis being given by Vonnegut." He claimed that ordinary atmospheric theories "cannot adequately account" for tornado wind speeds, which he believed were "close to Mach 1" (the speed of sound).

Could one snuff out tornadoes as one turns off a lightbulb—by switching off their sources of electricity? Colgate said that, to date, lightning suppression experiments with wires "have met with only partial success . . . presumably because the cloud itself is an insulating medium and will support large internal [electrical] fields without breaking down to the wires."

He suggested an alternate method: Inject tons of a negatively charged gas and "a fine aerosol of smoke particles" into the thundercloud. These might de-electrify the cloud before it could spawn tornadoes. For tornado suppression, ten tons of the gas might do the job. True, ten tons "is a large mass to inject rapidly adjacent to or inside the tornado, but is perhaps feasible from an airplane. Much larger quantities of borate are dropped on forest fires."

Colgate said he wouldn't advocate "bombing" every tornado that came along. "Bombing" was justified "in those rare cases where heroic measures may be justified to protect a city in the path of a major tornado. . . . A large city such as Chicago that may be threatened with major . . . destruction by a tornado may be more inclined to contemplate such heroic measures upon that rare occasion."

At about the same time, Vernon Rossow was working at NASA's Ames Research Center, just south of San Francisco. He was a bright forty-year-old expert on fluid dynamics and had just won NASA's "Super Performance Award." In 1966,

Time magazine wrote about his lab experiments to demonstrate links between electricity and tornadoes. He wanted to fire long wires into clouds to suppress lightning, thereby "robbing the tornado of the energy needed to sustain it," the magazine said.

By 1967, Rossow decided to conduct initial experiments on waterspouts. Waterspouts are weaker than tornadoes and, therefore, safer to work around. They are also more frequent than tornadoes, and therefore easier to locate. He headed for the waterspout-haunted Florida Keys, where he hoped "to fire wire-deploying projectiles into the cloud over any waterspout within range."

The age of tornado control was finally dawning. Or so it appeared. Even the world's most respected tornado expert took the topic seriously. To the news media, T. Theodore Fujita of the University of Chicago was "Mr. Tornado." In 1972 *National Geographic* ran a full-page color photo of Fujita playing with a miniature tornado—a swirling column of vapor—in his laboratory. Said Fujita: "I hope that within ten years we will learn from experiments like these how to modify real tornadoes."

> In [whirlwinds] passing over ponds or rivers, water is invariably raised in considerable quantity. This last remark is sufficient to show that waterspouts and tornadoes are essentially the same.
>
> —Elias Loomis (1842)

> Any attempts to modify a severe storm with potential or actual tornadoes obviously will have to be carried out with extreme caution, and it would be wise to experiment first with waterspouts, which are less of a threat. Actual modification attempts on menacing tornadoes are probably several years away.
>
> —Robert Davies-Jones and Edwin Kessler (1974)

The Florida Keys is a string of islands sprinkled through gorgeous blue-green waters south of the "Sunshine State." There were far worse places to be in the summer of 1967: Race riots were turning American cities into bonfires, and more U.S. troops were disappearing into the Vietnam quagmire. Meanwhile, Vernon Rossow was soaring over the Florida Keys on a romantic quest for waterspouts. If all went well, the NASA scientist would make history twice over: First, he would verify Bernard Vonnegut's theory that tornadoes (or waterspouts, at least) are electrically driven. Second, he would find a way to kill waterspouts and tornadoes.

Rossow and his NASA colleague, Harold Clements, had prestigious backers. Stanford University and the Navy gave them devices to measure electrical or magnetic activity near waterspouts. The Navy normally used its magnetic-anomaly detection (MAD) equipment to spot submerged Soviet submarines. The MAD gadgets were supersensitive: When Rossow and Clements' airplane flew 1,000 feet over a merchant ship, the MAD needle soared and didn't stabilize for a half-minute.

The Coast Guard allowed Rossow and Clements to base their waterspout-killing experiments on Coast Guard ships. For weeks, the two scientists sailed on three Coast Guard boats (Diligence, Ariadne, and Active) among the Keys, Cuba, and the Bahamas, looking for waterspouts. They carried small rockets that could tow long wires into the waterspout clouds.

Unfortunately, nature didn't cooperate. For five weeks, Rossow and Clements patrolled the warm seas, searching the horizon for waterspouts. They didn't want to fire until they were within range of a waterspout cloud. Sixteen times they saw waterspouts from the boats—white threads dangling from cumulus clouds. But they never came close enough to fire. Waterspouts are will-o'-the-wisps, like mirages; chase one, and it may disappear before you reach it.

Eventually they abandoned the waterspout-modification experiment. It was probably a waste of time anyway, for by then their MAD observations had undermined the experiment's key premise—that electricity powers waterspouts. On September 30, 1968, after flying near 52 waterspouts, Rossow concluded in his report to NASA that "electricity does not play a primary role in [waterspouts'] structure and could be eliminated as a generating mechanism or as a means of identification."

Indeed, atmospheric electricity seemed more likely to *destroy* waterspouts than to create them. On August 9, 1967, while gazing through binoculars, Rossow saw a lightning bolt near a waterspout funnel. The bolt "caused the funnel to break into pieces as if made of glass. . . . The pieces drifted apart slightly as they evaporated" and disappeared in a minute.

In his report, Rossow cautioned that he had been observing waterspouts, not tornadoes. Perhaps electricity triggered tornadoes but not waterspouts. Even so, his experiment marked the beginning of the decline of Vonnegut's electrical theory of tornadogenesis. The coup de grâce (as far as the scientific community was concerned) came several years later. By that time scientists from NSSL, the University of Oklahoma, and other institutions were routinely chasing tornadoes. In 1975, Robert Davies-Jones and Joseph H. Golden of NSSL wrote in the *Journal of Geophysical Research* that during their chases, they found no evidence to support Vonnegut's theory. Chasers rarely saw lightning near tornadoes. They also surveyed tornado tracks, checking for evidence of unusual electrical activity—say, scorched ground. They found none.

Also by that time, they noted, one of Vonnegut's central assumptions—that tornadic winds approach the speed of sound—had collapsed. Scientists had analyzed motion-picture films of tornadoes to estimate how fast they were rotating. They also surveyed building damage to estimate what wind speeds would create such wreckage. Their conclusion: Twister winds never exceeded 300 mph, and most are less than 200 mph. Tornadoes weren't as insanely energetic as previously thought. They were still incredibly energetic—too energetic for theorists to fully explain (even today). But the lowered wind speeds persuaded theorists that their task wasn't as daunting as it once seemed, and that far-out theories like Vonnegut's were probably unnecessary.

Davies-Jones and Golden had thrown down the gauntlet, and Vonnegut and Colgate responded feistily. Vonnegut argued that the chasers might have seen more electrical activity if they chased tornadoes at night. Also, perhaps they hadn't searched as "diligently" as they should for scorched ground.

Davies-Jones and Golden fired back: They *had* chased tornadoes at night; they *had* searched diligently for scorched ground.

Backed into a corner, Vonnegut responded philosophically: "Even if it eventually proves possible to provide a satisfactory explanation for severe tornadoes that is based on traditional meteorological mechanisms, it will not exclude the possibility that electricity could in some cases play a significant role." To some, Vonnegut was saying that it was no longer his responsibility to prove himself right; it was his critics' job to prove him wrong. Davies-Jones and Golden replied scornfully: "Vonnegut has adopted a very secure defensive position. His last paragraph indicates that it is practically impossible to prove him wrong." After that testy exchange, talk of electrical generation of tornadoes virtually disappeared from scientific journals. Tornado experts frequently cite the Davies-Jones-Golden paper as the one that nixed the debate over electricity and tornadoes.

Vonnegut hasn't given up. In 1990 the elderly, amiable scientist flew to Plainfield, Illinois, to interview and videotape eyewitnesses who reported seeing eerie electrical effects during a tornado. However, almost all tornado experts agree that Vonnegut's theory is a dead duck. By the mid-1990s, hundreds of storm chasers had videotaped tornadoes at close range; the tapes show little, if any, unusual lightning activity close to funnels. If there's any link between tornadoes and atmospheric electricity, then the link must be very subtle indeed.

The demise of the electrical theory didn't kill hopes for tornado control—not by a long shot. In fact, the topic was discussed vigorously in the late 1960s and early 1970s. Most scientists had never accepted Vonnegut's electrical theory anyway, so its death didn't faze them. They speculated about other, nonelectrical schemes for subduing tornadoes.

Gordon J. F. MacDonald of UCLA, a member of President Johnson's Science Advisory Committee, thought cloud seeding offered possibilities for tornado modification. True, "the typical amount of energy expended in a single tornado is equivalent to about fifty kilotons of explosives," that is, two and a half Hiroshima-type atomic bombs. ". . . These vast quantities of energy make it unlikely that brute-force techniques will lead to sensible weather modification. Results could be achieved, however, by working on the instabilities in the atmosphere"—that is, using cloud seeding or other means to trigger small changes in weather that would cascade into larger effects that might, with any luck, weaken or suffocate a tornado.

In 1972, an article titled "On Tornadoes and Their Modification" appeared in MIT's prestigious journal *Technology Review*. The author was NSSL director Edwin Kessler. He cautioned that tornado modification remained in the realm of speculation. However, he discussed several possible techniques at length. One strategy required the generation of "hot spots"—areas of artificial convection—that might destabilize or block tornadoes. Scientists could generate a hot spot by affixing jet engines to the ground and pointing them upward. They would exhaust "half a ton of air per second at a speed of 1,000 ft./sec.," thereby creating a hot updraft. The updraft could spawn local clouds and perhaps precipitation, which might suck energy from another storm that threatened to turn tornadic. (Alternately, the rain might create a cold downdraft that might weaken the tornado.) However, the plan could backfire: The jet engines might themselves spawn a severe storm!

Scientists might also form a hot spot by burning large containers of oil or coal. The French researcher Henri Dessens had already built such a facility—called "Meteotron"—in

France. He burned enormous amounts of fuel in hopes (never realized) of generating significant amounts of rainfall. (Ironically, the blaze created small tornadoes!)

Another possibility, Kessler said, was to "alter the Earth's topography and roughness so as to decrease the probability of tornadoes over inhabited areas—perhaps by building special-purpose mounds or ridges or by planting wind-resisting vegetation." Scientists had long debated whether terrain affects tornadoes. Long ago, they thought that minor topographical features (such as hills) could divert twisters. They no longer believe this (nor do the residents of Barneveld, Wisconsin, and Topeka, Kansas). But laboratory experiments and field observations hint that "rough" terrains such as forests or buildings in large cities may weaken tornadoes. If so, then one might protect a city by, say, planting a forest around it.

Others advocated a more aggressive approach. In the late 1960s and 1970s, a few researchers used laboratory models of tornadoes to investigate ways to destroy real-life twisters with explosives. An aerospace engineer, T. Maxworthy of the University of Southern California, suggested testing the scheme by detonating explosives inside small desert vortices called dust devils. He added lightheartedly: "The legal aspects of such a field experiment are interesting to contemplate!"

Would-be tornado controllers weren't merely daydreaming. They were motivated by a fear of future disasters on an unparalleled scale. That fear was worsened by the events of April 3 and 4, 1974, when the nation experienced its worst recorded tornado outbreak. An incredible 30 tornadoes ranking F4 or F5, plus many more of weaker size (the total number of tornadoes may have approached 150) raked the Eastern United States. Their roughly parallel paths ran across a map like scratches from a gigantic cat's claw. More than 300 people died, 34 of them in a single town—Xenia, Ohio.

The 1974 cataclysm especially troubled officials in a then-thriving U.S. business: the nuclear power industry.

The proliferating nuclear power industry and massive destruction brought by the April 3–4, 1974 tornado outbreak have brought into sharper focus the problem of tornado structure and behavior.

—Joseph Golden (January 1976)

The 1970s was a nervous time for executives of the nuclear industry. On the one hand, their prospects looked bright. Shaken by the Middle Eastern-OPEC oil embargo of 1973, the United States and other advanced industrial nations planned a big shift to alternate energy sources, such as nuclear power. Hundreds of nuclear power plants were on the drawing boards. A nuclear power reactor generates energy by splitting atoms (fission), but much more slowly than in an atomic bomb. A reactor can power a town for weeks with energy extracted from a few pellets of nuclear fuel, small enough to hold in your hand.

On the other hand, an antinuclear movement was on the rise. Critics, including some scientists, warned that a reactor might overheat and explode, ejecting wastes that would remain highly radioactive for centuries. Experts debated possible causes of a reactor accident, including terrorist attacks, earthquakes, floods, hurricanes, airplane crashes—and tornadoes.

What would happen if a twister hit a nuclear plant? Were the plants sturdy enough to withstand the tornado's fierce winds? And if not, would the tornado spew poisonous, scalding-hot radioactivity over the countryside? Could a tornado unleash a nuclear catastrophe, rendering much of a state uninhabitable? To find answers, the U.S. Atomic Energy Commission and its successor, the Nuclear Regulatory Commission or NRC, funded extensive research on tornadoes from the 1960s into the 1980s. Specifically, the NRC needed to know: Are tornadoes violent enough and frequent enough to threaten commercial nuclear reactors? And if so, how can nuclear engineers redesign reactors to make them twister-proof? The key questions were:

(1) How fast are a tornado's fastest winds?

The pressure exerted by wind rises by the square of its speed increase. For example, if a wind doubles in speed, then its pressure quadruples (2 times 2, that is, 2 squared); if it triples in speed, then its pressure soars by nine times (3 squared); and so on. So it made a very big difference whether tornadoes' maximum winds were 300 or 400 or (as some analysts thought) more than 500 mph. The costs of safeguarding a structure against, say, 400 mph winds would be formidably higher than for 200 mph winds—perhaps too high for a utility company to justify buying a nuclear reactor.

(2) What is the internal "structure" of a tornado? Is it what it appears to be from a distance—a simple vortex, into which air flows evenly from all directions, then ascends upward into the thundercloud? Or is the funnel more complex or turbulent than that? For example, does air flow in different directions and at different speeds in different parts of the twister? Engineers needed answers to such questions so they could devise realistic computer simulations of a tornado's impact on a nuclear plant.

(3) How *frequent* are severe tornadoes? This question was probably the most important one—and the hardest to answer.

Tornado "climatologists" study tornado frequencies in different regions. They rely heavily on historical records such as old newspapers. Unfortunately, newspaper records are often incomplete and inaccurate. They rarely contain enough detail for scientists to determine whether a "tornado" in, say, 1880, was actually several "tornadoes" seen in different locales. To patch the holes in tornado records, the agency hired researchers such as Thomas P. Grazulis to scour old newspaper files and archives around the country for information on old twisters.

The NRC needed answers to these and other questions fairly quickly. By the 1970s, nuclear power was controversial; its base of political support was fragile. A single devastating accident could wreck the industry.

On April 18, 1978, Robert F. Abbey Jr. of the NRC called Ted Fujita, a.k.a. "Mr. Tornado." Abbey asked Fujita to fly to Mississippi and do something that Fujita had done many times before—survey a tornado's damage path. But the Mississippi tornado was unique: It had just hit a nuclear power plant.

Fortunately, the plant's reactor hadn't been operational at the time. The facility was still under construction. But what if it had been operational? Would the plant have withstood the blow? Or would the reactor containment vessel have ruptured? Fujita hoped to find out.

On reaching Mississippi, Fujita flew in a small airplane over the tornado's path. He snapped numerous photos as the Cessna 182 soared over the 18-mile trail of debris and destruction, which climaxed at the half-completed Grand Gulf nuclear power plant near Port Gibson. Aerial surveys were one of Fujita's specialties. He had flown over and photographed hundreds of tornado tracks. As an autopsy reveals a person's cause of death, a tornado path reveals much about the storm's structure and dynamics. By carefully analyzing the type and distribution of debris, scientists could begin to answer questions (1) and (2) on the NRC's list. Fujita's answers to these questions—especially (2), regarding tornado structure—would make him famous.

The Mississippi twister was a spinoff of severe storms that had dropped ten tornadoes on Arkansas, Louisiana, and Mississippi. By 11:00 P.M. on April 17, the thunderstorms crossed the Mississippi River and approached the power plant. The Grand Gulf plant was one of Mississippi's proudest construction projects. When finished, planners promised, its twin reactors would generate 2.5 gigawatts and bring the blessings of cheap electricity to the impoverished bayous.

Joe Pool was the construction superintendent for the night shift. He hadn't received a tornado warning. But he knew better than to risk losing staff and equipment in high winds. Huge cranes—which resembled giant, metallic praying mantises—loomed over the still-unfinished nuclear reactor containment building. One big puff of wind, and over they would go. Pool ordered his crane operators to secure their gear. They did, then glanced at the lightning-lit sky and scurried indoors. The tornado roared by at 11:30 P.M.

Fujita could tell what happened next, as he flew hundreds of feet over the wreckage. First the tornado shoved a heavy mobile device called a sand spreader 300 feet into a grassy field. Then it whacked the cooling tower, a tall, carafe-shaped structure. Within the tower, a crane collapsed and knocked out a big chunk of the tower's upper wall. On the ground, barrels scattered like windblown confetti. Light poles snapped like dry pasta. Workers hid inside the plant and listened to the crashing and clattering outdoors.

Then the tornado advanced on the plant's crown jewel: the reactor containment building. Containment Unit One resembled a giant, open bowl; its roof hadn't been installed. By this time the twister had weakened from an F2 to a mere F1, but it was still fierce enough to topple another crane. The crane weaved drunkenly, then collapsed with a screech into the containment building.

The tornado surged on, toppling piles of bricks and shredding a chain-link fence. Then it rumbled off into the night.

In short, the plant was a mess—not a debacle, exactly, but not good news for nuclear designers, either. If an F1 tornado could do this kind of damage, then what might an F5 do? What would have happened if the plant had been operational and an F5 tornado had picked up (say) the sand spreader and hurled it at the containment dome? Would the dome have held? Or cracked like a china bowl?

During Fujita's flight he passed over Lake Brun, an "oxbow" lake—so called because of its crescent shape. Near Lake Brun was a flattened forest. Fujita studied the orientation of the fallen trees. Until recent years, he would have assumed that the tornado had knocked the trees down. But if that were the case, then all the trees should have fallen in the same direction while the twister, moving horizontally over the ground, mowed them down. Instead, the trees spread out radially (in all directions), as if hit by an immense force from *above*.

To Fujita, the radially toppled trees were an increasingly familiar sight. He had seen similar damage on numerous flights. By the late 1970s, he claimed he had discovered the cause: a "downburst" or "microburst," a surge of cold air that plunged from a thunderstorm. The downburst fell like a mailed fist, squashing objects directly beneath it and spreading them out radially. (Likewise, an egg dropped vertically to the floor splatters out in all directions.) Fujita claimed that some "tornado" damage was actually caused by downbursts. More important, he believed that downbursts sometimes strike planes, causing mysterious aviation accidents. Originally controversial, his downburst theory is now generally accepted.

Fujita got the idea for downbursts from a similar sight he had seen at a very different time and place: Hiroshima.

September 1945: Twenty-four-year-old Japanese physicist Tetsuya Fujita flew in a plane over the still-radioactive rubble of Hiroshima. A month before, an American bomber had hit the city with an atomic bomb. Fujita noticed that some debris spread out radially. The explanation was obvious: The bomb must have exploded not on the ground but, rather, overhead. When Fujita and his associates landed, they surveyed the damage from a hillside cemetery. Burned bodies still littered the hill. They found that when the bomb detonated, it burned the inner rims of bamboo flowerpots in the cemetery, except those parts in shadow. The unburned areas were still visible— ghostly reminders of atomic hellfire. By measuring the shadows, they calculated the altitude of the explosion: 1,700 feet. The bomb had, indeed, detonated overhead. That explained the radial distribution of debris.

The radioactive ruins of Hiroshima and a toppled forest in Mississippi had, it seemed, little in common. But Fujita's ability to perceive links between seemingly unrelated tragedies inspired his discovery of downbursts. It's an example of the scientific imagination at its best. That same imagination enabled him to make his greatest discovery—"suction vortices"—and to give scientists their first close look at the wild, weird world within tornadoes.

He was born Tetsuya Fujita on October 23, 1920, at Kyushu, a Japanese island "adorned with two active volcanoes. I visited these volcanoes during eruptions, observing the angry face of our living planet," he recalls in his autobiography. He was slightly irreverent. In middle school, a teacher taught Tetsuya about a venerable monk who spent 30 years digging a tunnel with a hammer and chisel. Tetsuya scoffed that if *he* had been the monk, he would have built a "digging machine" to shorten the work. "I did not receive a passing grade," he notes.

He studied mechanical engineering at Meiji College, then became an assistant physics professor there. In late 1941, the Japanese invaded Pearl Harbor and America entered the Second World War. Late in the war, as U.S. bombers pummeled the Japanese home islands, the Japanese Navy asked him to find a way to use searchlight beams to detect enemy aircraft. The atmosphere bends light, like a prism, and he had to take this fact into account during his research. The project steered him toward meteorology. After the war, in September 1948, he and his fiancée, Tatsuko, visited the site of a twister on his home island. "We were shocked to see roofs blown off residential houses and rice crops flattened in rice fields." He mapped the twister's 6-mile path. A colleague noticed an English-language article on thunderstorms in a garbage can and passed it on to Fujita. Fujita read the article and, typing one key at a time on a portable typewriter, wrote to the author—Horace Byers, a leading thunderstorm researcher at the University of Chicago. Impressed by Fujita's research, Byers brought him to the university as a visiting research associate.

One of Fujita's major tornado studies occurred in 1957. On June 20, a major twister passed near Fargo, North Dakota. Drivers on Highway 10 spotted it—a huge cloud on the ground—swirling toward the city at 20 mph. When they got to Fargo, they reported the news to police. Word spread fast. Soon radio and TV stations were warning locals to run for shelter. Most did, but some grabbed their cameras and stood outside and scanned the darkening horizon. The whirlwind surged into town, killing 10 people, injuring more than 100, and damaging or wrecking 1,300 homes.

Fujita contacted Dewey Bergquist, a weather forecaster with WDAY-TV in Fargo, and asked him: Would he announce on the air that the University of Chicago needed photos of the twister? Bergquist did. The result was a photographic deluge—150 photos and 5 home-movie films. Fujita analyzed the movie films frame by frame. By this tedious process, called photogrammetry, he estimated how fast the tornado moved and how quickly its winds rotated. He published his study of the Fargo tornado in 1960. More than three decades later it remains a classic—"arguably, the single most comprehensive set of eyewitness photographs and accounts" of any tornado, as Howard Bluestein and Joseph Golden observed in 1993.

Fujita first surveyed a tornado track from an airplane in 1965. Over the next twenty-six years, he and his associates flew over more than 300 tracks and snapped some 30,000 photos. During his early flights, he noticed strange patterns in the cornfields. The patterns resembled a repeating sequence of letter "C"s that slightly overlapped.

A decade earlier, the researcher E. L. Van Tassel had noticed similar "cycloidal" marks in tornado tracks. Van Tassel proposed the marks were huge scratches, gouged out of the ground by debris swirling in the tornado. He calculated the debris could gouge out the marks if the winds were blowing close to 500 mph.

On April 11, 1965, one of the harshest tornado outbreaks in history hit the United States. Thirty-six twisters attacked six states—Iowa, Indiana, Illinois, Wisconsin, Michigan, and Ohio. Fujita rented a Cessna and spent a week flying over

7,500 miles of tornado paths, a distance equaling two-and-a-half-times the diameter of the United States. Later he scrutinized his photos of a tornado track in Kokomo, Indiana. The track contained distinct cycloidal markings. As he studied the marks, he became convinced that they weren't scratches. Rather, they were tornado debris—fragments of buildings and trees and other junk that the tornado had swept into odd, looping piles. In 1967 Fujita visited a tornado trail in Illinois on the ground and got a close-up look at cycloidal marks. Sure enough, they weren't scratches; they were debris. How could a tornado—a simple vortex—arrange debris into such neat geometric patterns?

The answer: A tornado *isn't* always a simple, spiraling vortex! Rather, it may include a number of smaller vortices, which orbit the larger, central vortex like satellites orbiting the Earth. Fujita suspected the tornado's strongest winds were concentrated at the small vortices, which he called "suction vortices."

Suction vortices were a pivotal discovery for at least two reasons. For one thing, they showed that tornadoes weren't simple structures; they were a complex structure of swirls *within* swirls. By far the fastest winds were centered in the suction vortices. The main funnel blows objects along, but the suction vortex *lifts* them. However, the suction vortex—like a powerful but inefficient vacuum cleaner—"cannot pick up everything it has collected, [so] a line of deposit is usually left behind its path." These debris deposits were cycloidal (the repeating "C" shapes) because the tornado's motion isn't a simple spiraling movement but, rather, a combination of *several* separate movements: (1) the rotation of each suction vortex around its own axis; (2) the revolution of the suction vortices around the main funnel; (3) the rotation of the main axis; and (4) the overall tornado's horizontal movement ("translation") across the terrain. Merge those four motions and you get the looping debris trails.

The discovery of suction vortices had important implications for the NRC research. The bad news was that suction vortices were extraordinarily violent, by far the windiest part of the funnel, but they were restricted to small parts of the tornado. Their small size and complex motion made it hard to figure out how wind stresses would be distributed around a nuclear plant. But that was also the good news: Only small parts of the tornado contained the worst winds, not the entire funnel. That lessened the chance that the highest winds would strike a sensitive structure. Based on his new interpretation of the cycloidal marks, Fujita concluded that Van Tassel had seriously overestimated the tornadic wind speeds; they were substantially lower than 500 mph. That, too, was good news for nuclear plant designers.

Tornado "pranks" are legendary. Countless American legends tell of twisters that ripped apart a well-built home but ignored a nearby shack, or that leveled the home but left a goldfish bowl standing in the living room. How can such events happen? Because the suction vortices are extremely selective: They concentrate the highest winds in extremely small areas.

Fujita made one of his most remarkable and grisly observations at Lubbock, Texas, in 1970, after two tornadoes smashed the town. He showed that the vast majority of deaths occurred precisely where the suction vortices touched down—that is, along the cycloidal pathways. The "random destruction" of tornadoes is a myth; their choreography obeys a lethal geometry.

Lubbock was a city of 170,000 with a passion for football and little experience with twisters. The last tornado had struck the town in 1900. The inhabitants—like the residents of many moderate-sized urban areas—assumed "it won't happen here." At 8:10 P.M. on the evening of May 11, 1970, the first, weaker twister arrived. It passed over an unfinished highway interchange and blew 13 beams off the overpass. Each beam weighed more than 50 tons. At 9:45 a fiercer tornado buzz-sawed from downtown Lubbock to the airport.

When Fujita arrived in town, he observed that the first tornado had knocked down only 13 of the 35 available beams.

Why did it leave the other 22 standing? Simple: because the cycloidal paths of the suction vortices took them over the 13 beams. They missed the other 22. To prove his case, he mapped the cycloidal debris paths with his usual meticulousness.

The twisters hurled a freight car more than 200 feet from its track, and blew a 16-ton empty fertilizer tank a half mile. Fujita concluded the huge tank had been airborne for two-thirds of its trip (at one point it passed over a four-lane federal highway). A "small wooden shack in a direct path of the second tornado was found to be practically undamaged because it was standing between suction [vortices]."

Later Fujita flew in a helicopter over the second tornado path. He observed debris marking the cycloidal paths of the suction vortices. He mapped the cycloidal trails. Then he marked the site of every fatality on the same map. To his amazement, *95 percent* of all fatalities occurred where the suction vortices had struck. In a macabre touch, his map describes the manner of each victim's death: for example, "woman died after being swept out of house and wrapped in sheet metal," "boy sucked out of car stopped at traffic light, killed by flying debris," "woman crushed to death under pickup tossed by storm," "family of 5 died in house, tossed 210 ft. southeast into field," and so on.

The "Fujita Scale" is based substantially on Fujita's Lubbock research. The scale was a big step toward a long-sought tool: a way to classify tornado strengths quantitatively, just as astronomers classify stars according to their specific masses, colors, and other traits. Tornado investigators use the Fujita Scale to estimate the severity of a tornado based on how much damage it causes. Tornadoes that are ranked F0 are called "gale tornadoes": They break off tree branches and topple trees with shallow roots. An F1 or "moderate" tornado may overturn a mobile home and damage roofs. An F2 ("significant") tornado can uproot a big tree and wreck a mobile home. An F3 ("severe") tornado may tip over a train and throw a car off the road. An F4 ("devastating") tornado can destroy a well-built home and toss cars through the air. An F5 ("incredible," such as the Barneveld tornado) twister can damage steel-reinforced concrete buildings and hurl missiles as big as cars for hundreds of feet.

In the 1970s, NSSFC director Allen Pearson modified the scale by adding specifics on the path length and width of each tornado type. The modified scale's categories are:

THE FUJITA-PEARSON SCALE

Scale	Wind Speed	Path Length	Path Width
0	40–72 mph	0.3–0.9 miles	6–17 yards
1	73–112 mph	1.0–3.1 miles	18–55 yards
2	113–157 mph	3.2–9.9 miles	56–175 yards
3	158–206 mph	10–31 miles	176–566 yards
4	207–260 mph	32–99 miles	0.3–0.9 miles
5	261–318 mph	100–315 miles	1.0–3.1 miles

The Fujita Scale is now widely used by tornado investigators. However, some prominent scientists feel the scale is inaccurate because it relies on subjective judgment. What looks like F2 damage to one scientist may look like F3 damage to another. Considering that F2 and F3 winds range from 113 mph to 206 mph (almost a 100 percent difference!), this subjective factor is a serious flaw. Damage is also hard to assess because of variations in the quality of workmanship and building materials.

Worse, the scale may inadvertently underestimate wind speeds. Rasmussen complains that structural damage "only tells you winds were at least strong enough to *destroy* a structure. They might have been much stronger." He suspects that several tornadoes have been rated F1 that were actually F3s.

The Fujita Scale is almost useless for estimating tornado wind speeds in rural terrain because there are few or no buildings to damage. Pearson himself has joked: "Having gone through the North Dakota [tornado] data base, [I know that] if it weren't for outdoor privies, no one would know what events occurred there."

Initially, some meteorologists didn't accept Fujita's theory of suction vortices. The idea gradually won acceptance for several reasons: First, researchers simulated suction vortices in miniature "tornadoes" in the laboratory. Second, they observed suction vortex-like objects in dust devils, which are smaller versions of tornadoes. And third, they found eyewitness and photographic evidence of suction vortices in past tornadoes—some of them more than a century ago!

The so-called "tornado machine" is one of the most useful and entertaining tools in tornado science. These laboratory devices use whirling fans and smoke generators to create small whirlwinds, typically several feet high. The whirlwinds are often amazingly similar to real-life tornadoes and waterspouts. Scientists analyze how air moves through the mini-twisters by spraying particles into them—say, confetti, sawdust, or soap bubbles. They measure air pressure inside the funnels by sliding small instruments into the vortices, or by scanning them with a laser beam.

Scientists use tornado machines to explore questions such as: Why do tornadoes come in so many shapes and sizes? Why are some tornadoes smooth, "finger"-like funnels while others are boiling, turbulent maelstroms? What causes suction vortices? Is a tornado one big updraft, or does it have a narrow downdraft in its core? How is a tornado affected by the terrain it passes over? How reliable are computer simulations of tornadoes? A few have used the machines to probe the boldest question of all: Is there any way to modify or "kill" a tornado?

One of the biggest tornado machines is at Purdue University in West Lafayette, Indiana. John T. Snow built and operated this device in the 1970s and 1980s. The machine is a cylinder, 9 feet tall and 9 feet wide. At the top is a fan, which pumps air from the cylinder. At the bottom, a rotating screen stirs air into rotation. Vents emit swirling white vapor into the cylinder. As air ascends in the rotating "updraft," a vortex quickly forms.

In the early 1970s, the tornado researcher Neil Ward simulated suction vortices in a tornado machine. He inserted air pressure-measuring devices into the mini-suction vortices. Their air pressures were "much lower" than in the primary vortex. Ward's work convinced many scientists that Fujita's suction vortices were real.

Oddly, a few nonscientists had "discovered" suction vortices decades earlier, but their observations were lost to history. In 1876 a reporter for the _Chicago Tribune_ wrote about a tornado that passed through the Windy City, killing 2 and injuring 35. As it retreated to Lake Michigan, "it was then composed of 8 or 10 columns grouped together, all whirling around a central point." This is an unmistakable description of suction vortices. "Another such scene may never come in this generation," the writer added, "and it is to be regretted that the cylinder could not have been caught and pickled for scientific investigation."

The author-biologist Gunther Stent calls such observations "premature discoveries." On a number of occasions in the history of science, laypeople have recognized phenomena before scientists did. A classic example is meteorites or "shooting stars." A few ancient Greeks knew that rocks fell from the sky, but scientists didn't believe it until the nineteenth century.

Do tornadoes have "eyes"? Everyone has heard about the hurricane's "eye"—a calm area at the center of the cyclone. Inside the eye, winds may drop to zero and clouds may disappear. The sun or stars may come out. Ironically, the passage of a hurricane eye is dangerous because unwary people leave shelter, thinking the storm has passed. Within minutes the other side of the storm strikes. Those who left the shelter may die.

Likewise, a few people briefly trapped inside tornadoes observed that the interior winds were calm. In 1962, a tornado passed through Newton, Kansas; witnesses said the center of the vortex was quiet, but smaller funnels "were appearing all around town at different locations, moving in various directions." (Were those "funnels" suction vortices?) The veteran chaser Howard Bluestein has seen a few wall clouds with "eyelike" features. In one case in 1978, he drove under a wall cloud in Oklahoma "and noted a bright, hollow, cylindrical tube extending upward." (Is that the true explanation for Roy Hall's "brilliant cloud"?)

Hurricanes develop eyes because warm air high above the storm sinks downward. As it sinks, it evaporates clouds, creating the eye, a "hole" in the storm that is obvious on satellite photos. For many years, scientists suspected that something similar happens in tornadoes. Ward and other scientists used tornado machines to create micro-twisters, then tossed confetti and other particles into them to see how air flowed through them. Sure enough, they observed a slender column of sinking air within the vortices. Later, VORTEX radar scientists scanned tornadoes and concluded that air was sinking in their interior.

That sinking air may also explain suction vortices. As the air sinks, it encounters rising air. The result is similar to what happens when a crowd of people ascending a staircase meets a crowd in descent: turbulence, as everyone scrambles to get around everyone else. In tornadoes that turbulence is called vortex breakdown. Vortex breakdown allows the lower part of the tornado to disintegrate into several smaller, independent whirlwinds—the suction vortices.

And that leads us to a fascinating question: Are big cities immune to tornadoes?

A strange fact haunts the history of tornadoes: Certain major cities have experienced remarkably few twisters. Consider Chicago: Tornadoes have raged throughout the Chicago suburbs over the last century, but almost never penetrated the densely populated downtown. Why? Is it just a coincidence—a lucky break for the Windy City? Or is there something about Chicago and other large cities that *repels* tornadoes?

In the early 1970s, Fujita conducted an interesting experiment. Using his tornado machine, he simulated a tornado's attack on downtown Chicago. He placed small rocks at the base of the machine to simulate skyscrapers and a small pool of water to represent Lake Michigan. He placed heating wires under the rocks to mimic the city's heat. (Cities tend to be warmer than their surroundings because of machinery, pollution, human bodies, and reflectivity from artificial surfaces such as glass buildings.) When he switched on the "tornado," the misty funnel thrived until it hit the rocks. Then it weakened. It started to break up when he boosted the heat. Afterward, he speculated that tornadoes thrive on a steady diet of warm air. Perhaps big cities' rising warm air is too unstable to "feed" strong tornadoes.

Fujita also analyzed the history of tornadoes in Chicago and Tokyo. He found a "horseshoe"-shaped area of downtown Chicago that "appears to be tornado free during the past 20 years." He found the same effect in Tokyo, where the "tornado frequency appears to decrease toward the center of the city and toward the outer suburbs as well." He noted that Chicago's and Tokyo's populations were, respectively, 7 and 11 million people. Sticking his neck way out scientifically, he suggested that a city might possess a "threshold population" of 4 million for "effective tornado suppression."

Likewise, in 1977 C. R. Snider studied 244 Michigan tornadoes and concluded that large cities have disproportionately fewer tornadoes than small towns. The likelihood of a tornado is "approximately *inversely* proportional to the size of the city," he declared. "Large cities seem to resist tornado touchdowns."

And Derek M. Elsom and G. Terence Meaden of England have found that since 1830, Greater London has experienced "relatively few tornadoes . . . compared with the outer areas of the metropolis and the surrounding countryside. . . . For example, within a 10 km [kilometer] radius from the city of

London it is evident that no tornado has been reported for more than 50 years."

A few years after his initial "Chicago" experiment, Fujita seemed to back away from its most extreme implications. "Despite such experimental evidence, we are not certain if the heat generated in Chicago is sufficient to kill all tornadoes. Existence of the tornado-free area in Chicago might be just accidental. Until more research is done, we should not simply assume a false sense of tornado security in our city. A large, violent tornado might manage to smash through the Loop, damaging skyscrapers and causing showers of window glass onto the streets."

Indeed, one could make a strong case that Chicago's alleged immunity is an illusion—an illusion bred, like so many modern illusions, by the disregard of statistics.

Consider Fujita's historical review of Chicago twisters. Writing in 1973, he listed two tornadoes in central Chicago (1876 and 1920) from 1876 to 1976. According to Grazulis's *Significant Tornadoes*, the entire state of Illinois had almost 500 tornadoes over approximately the same period. Chicago covers 228 square miles, or 4/10ths of 1 percent of the total area of the state (57,918 square miles). Now suppose that tornadoes occurred randomly across the state of Illinois during that century. How many tornadoes would likely have hit Chicago or any other area of comparable size?

Simple arithmetic yields the answer: two! Exactly as many tornadoes as Chicago had experienced!

Therefore one could argue that Fujita's "big city" effect is indistinguishable from random chance—meaning it is a statistical illusion. If so, then Chicagoans are no more or less safe from tornadoes than much smaller communities.

Tornado climatologists have used computers to generate maps that show the paths of every major tornado in U.S. history. These maps show a thicket of lines, like scratches left by an army of cats. These maps are useful because they show that Chicago-sized tornado "gaps" exist all over the country. The gaps are scattered hither and yon, with no rhyme or reason. By the laws of probability, one expects tornado rates to vary from place to place because of simple natural randomness. So why isn't anyone talking about the *other* gaps? Probably because most of them are out in the middle of nowhere. It's easier to notice the so-called Chicago "gap" because it's in a big city. Chicago's tornado "immunity" may be simple luck—luck that may soon run out.

Or maybe not. Maybe there really is something about cities, some physical factor, that tends to repel tornadoes. Laboratory and field research suggests that tornadoes are affected by the terrain they pass over. This certainly happens with waterspouts: They almost always break up on hitting land.

Field observations of tornadoes tend to be contradictory, though. On the one hand, Japanese researchers report that a tornado intensified over smooth terrain and weakened over a community. On the other hand, in Indiana, a tornado grew more *violent* as it penetrated a grove of trees.

Suction vortices seem more likely to form over smooth terrain. When the Oshkosh, Wisconsin, tornado of April 21, 1974, passed through a field, it contained suction vortices; when it entered the town, the vortices melted into a single funnel.

As if things aren't complicated enough, consider this: Cities seem to *attract* ordinary (nontornadic) thunderstorms! In the 1970s S. A. Changnon studied the climate around St. Louis, Missouri. He concluded that the city and points immediately east had "more thunderstorms and hailstorms, more hail and lightning strokes per unit area when they occur, longer lasting periods of thunder activity and hail, and more frequent high winds and damaging hail." Similarly, researchers have noted an unusual amount of thunderstorm activity over London. Clearly something is wrong here: How can a city attract thunderstorms but repel tornadoes?

Clearly, we have a great deal to learn about the workings of our atmosphere. So for the time being, residents of big cities shouldn't get too smug; just because they're surrounded by skyscrapers doesn't necessarily mean they're safe from twisters. For the foreseeable future, a wise city resident who hears the sirens sound or sees a "tornado watch" on TV should do what a Kansas farmer would do: Head for shelter. Don't risk your life based on a few scientists' laboratory experiments with rocks and ceiling fans.

The NRC continued to fund tornado research into the 1980s. Then the cash dried up. Why? The reasons remain a little mysterious. Perhaps it's because of the election of a new President, Ronald Reagan. He came from California, a state that cared more about earthquakes than tornadoes.

Or perhaps it's because the nuclear industry decided that tornadoes weren't a big threat after all. In the mid-1970s, the NRC funded research at Sandia National Laboratories in New Mexico, where engineers conducted a dramatic experiment. They placed a telephone pole on a rocket sled and fired it at hundreds of miles per hour at a steel-reinforced concrete barrier. The barrier resembled the containment building around a commercial reactor. The telephone pole slammed into the wall and disintegrated; the wall survived with hardly a scratch.

Or perhaps NRC officials lost interest in tornadoes when, ironically, their oldest fear came true: a major nuclear accident. On March 28, 1979, a commercial nuclear reactor at Three Mile Island near Harrisburg, Pennsylvania, suffered a partial meltdown. Total damages exceeded a billion dollars. The accident wasn't caused by a tornado—or by terrorists, or a flood, or a hurricane, or an airplane crash, or a falling meteor. It was caused by human error—by well-meaning but inadequately trained personnel who pressed the wrong buttons at the wrong times. The accident spurred many utility companies to cancel their contracts to build nuclear reactors. If the nuclear industry continues to stagnate, then a severe tornado will probably never hit a nuclear reactor because there will be hardly any to hit.

However controversial some of his conclusions, Fujita's research revolutionized scientists' understanding of the inner world of tornadoes. Twisters, as it turned out, had an elegant inner geometry that varied from moment to moment during their lives. That geometry caused their wind speeds to vary from place to place within the funnel. And that fact, in turn, greatly complicated the task of determining their likely impact on buildings, from ordinary homes to nuclear power plants.

But a crucial question lingered: How did tornadoes acquire their vicious energy? To answer that question, it wasn't enough to study films of tornadoes, or to survey their debris paths after they had passed. Rather, it was necessary to penetrate the tornado funnel itself—to pierce it with an instrumented probe that would radio back information about the interior. That task—the scientific world's version of Dorothy and Toto's flight on the whirlwind—would prove much more difficult and dangerous than expected.

CHAPTER 5

Into the Whirlwind

I heartily endorse the . . . plea for more measurements of the electrical state very close to the tornado funnel. Perhaps some fortunate observer can arrange to place his measuring apparatus inside the funnel and still emerge with useable records.

—Ross Gunn, *Journal of Meteorology* (June 1957)

Humans have always ventured into forbidden places. They've spelunked in deep caves, ridden submarines to the ocean's darkest depths, and defied superstitious edicts by penetrating the crypt of Tutankhamen. Likewise, people have dreamed of venturing into a tornado. Many have, usually unwittingly. Only a few lived to talk about it.

In 1899, a twister hit Kirksville, Missouri, and blew two women and a boy over a church. They landed more than a thousand feet away, unharmed. "I was conscious all the time I was flying through the air . . ." one woman later said. "I seemed to be lifted up and whirled round and round, going up to a great height, at one time far above the church steeples. . . . As I was going through the air, being whirled about at the sport of the storm, I saw a horse soaring and rotating about with me. It was a white horse and had a harness on. By the way it kicked and struggled as it was hurled about I knew it was live. I prayed God that the horse might not come in contact with me, and it did not."

Roy Bennett was a sergeant at Sheppard Air Force Base outside Wichita Falls, Texas, in April 1964, when a twister smashed his home. He later told the *Houston Post* how he clung to basement pipes until the gale snatched him skyward. "I was circling upward, inside the tornado's funnel." A tractor and a bed flew past him. In his dazed state, he gazed at the bed and thought: "If I could just get over to that bed, I could take a nap." He looked down at the town—a few hundred feet below—as the whirlwind ripped his community to pieces. Later, emergency teams found him lying 300 feet from his home. He was tangled in barbed wire and almost dead. He spent nearly two years recovering in a hospital.

Of course, a twister is only the fierce lower extension of a grander maelstrom: the thunderstorm. Until the 1940s, only a foolhardy pilot ventured near a thunderstorm. Its extreme updrafts and downdrafts could slap a plane from the sky. In the late 1940s, though, pilots deliberately hauled scientific instruments through thunderstorms as part of meteorologist Horace Byers's so-called Thunderstorm Project in Ohio and Florida.

In 1959, by accident, Marine Lieutenant Colonel William H. Rankin got a close-up view of a thunderstorm's inner hell. On July 26, he took off in an F8U Crusader jet fighter from a North Carolina military base, bound for Massachusetts. He wasn't wearing a pressure suit—only "ordinary pilots' coveralls." Seeing thunderstorms in Virginia, he tried to fly over them. Then his engine's red warning light flickered. Fearing an explosion, he ejected nine miles above the Earth. The outside air was painfully cold—70 degrees below zero. In the rarefied atmosphere, air rushed out of his body; his abdomen swelled "as though I were in well-advanced pregnancy" and blood seeped from his eyes, ears, nose, and mouth. He plunged through the "white wall" of the thundercloud and into the tempest. Blue sheets of lightning exploded around him. His chute automatically opened. Then began his roller-coaster ride through the storm, as updrafts and downdrafts repeatedly flung him up and down. Intense rain almost drowned him, "as though I were at the bottom of a swimming pool." He repeatedly vomited. Baseball-sized hail battered him senseless. At one point he stared "down into a long, black tunnel, a nightmarish corridor in space." (A mesocyclone? An embryonic tornado funnel?) "Sometimes, not wanting to see what was going on, I shut my eyes."

After forty minutes he landed ignominiously: He crashed into a tree trunk. A passing motorist gave him a ride. After a hospital stay—which included a brief bout of mild amnesia—he returned to active duty. He later described his adventure in an article for the *Saturday Evening Post* and a book, *The Man Who Rode the Thunder*.

Even more incredibly, in 1963 an entire commercial DC-8 jetliner may have flown through a tornado funnel! On the late morning of November 9, 1963, Eastern Air Lines flight 301 left New York City, bound for Mexico City. It carried 128 passengers. The pilots were Captain N. H. French and First Officer Grant Newby. The plane encountered thunderstorms over the Gulf of Mexico and stopped in Houston for refueling. Then it took off again. French planned to lift the plane to a cruising altitude of 31,000 feet. At 18,000 feet the plane passed through a cloud layer. French later testified at a Civil Aeronautics Board (CAB) hearing that just before entering the clouds,

I could see visually this heavy dark area to my left. On the right I could see visually a small dark area. In other words, it looked like a [thunderstorm] cell over on the right. However, my airplane radar did not show any echo for the cell on the right-hand side, or to the north. The radar did show the heavy band of dark area on the south and to my left. So on this 270-degree heading, we continued to climb and continued on up. There was a third cloud formation and the cloud formation was in the form of a sort of arch, that appeared to be an arch between a cell on the right and a cell on the left.

At exactly 20 seconds after 3:00 P.M., flight 301 entered the cloud layer and began to encounter turbulence. Light hail rattled against the plane. Newby switched on the "Seat Belt" light in the passenger section.

At 19,300 feet, something went terribly awry. French, Newby, and the flight engineer were watching two different air speed monitors. The monitors indicated their speed had dropped to *zero*. They stared at the dials for a moment, dumbfounded. Suddenly, the floor fell out from under them and the plane went into a steep dive. They couldn't believe their eyes.

Back in the passenger cabin, numerous passengers hurtled from their chairs. "I was just glued to the ceiling," Robert L. Monahan of Ocean City, New Jersey, later said.

Faster and faster the DC-8 plunged, its speed accelerating to an astonishing nine-tenths of the speed of sound. Newby wrestled frantically with the flight gear, struggling to save his stricken plane. He pushed the engines into idle reverse thrust. This maneuver (which tornado analyst Ferdinand C. Bates later called "one of the most amazing maneuvers of modern aviation") may have rescued the airplane. The DC-8 slowed down and gradually lifted.

Twelve seconds later, Newby heard a loud SNAP! The plane's number three jet engine had broken off and hurtled to Earth.

It took the crew two minutes to regain full control of the plane, using their three surviving jet engines. They had fallen 14,000 feet—almost 3 miles—before recovering. The plane circled Barksdale Air Force Base, dumping fuel, then safely landed there. Seventeen passengers were injured, but everyone survived what could have been a disaster. The Civil Aeronautics Board launched an investigation of the near-tragedy.

During the hearing, French made an incredible claim: His DC-8 had flown into a tornado. He insisted there was no other explanation for his plane's extraordinary behavior. Unfortunately, he assumed that a tornado would be a vertical column. This assumption hampered investigators' effort to understand certain aspects of the mishap as they reviewed it, second by second. In the end, the CAB blamed the accident on a technical detail: "The deterioration of aircraft stability characteristics in turbulence [probably occurred] because of abnormal positioning of longitudinal trim components."

Several years later, Bates, a researcher at St. Louis University, concluded that French was right: He had flown into a tornado funnel. Puzzling aspects of the flight made sense, Bates said, if one made an unusual assumption: The tornado was inclined to the horizontal. He stressed that the DC-8's air speed had plunged to zero. *"There is no maneuver in which the indicated air speed of a DC-8 can be brought to zero in flight in a no-wind environment. . . .* The *only* way such an air-speed decrease can take place is through the penetration of a gradient of tail-to-nose wind so intense that the aircraft cannot initially respond. . . . The only phenomenon we know which contains such a wind speed and gradient," Bates said, "is the tornado." [His emphasis.]

Since the 1960s we have learned about another hazard to aviation—Fujita's downbursts (a small, high-speed downdraft). Could Newby have run into a downburst and assumed it was a tornado?

Possibly, but the case for a tornado remains intriguing. Scientists now believe that mesocyclones generate many severe tornadoes. Mesocyclones start as horizontal, spinning cylinders of air, which updrafts push into vertical positions. Meteorological journals have run photographs of horizontal funnels. One appears in the March 1978 *Bulletin of the American Meteorological Society*. It resembles a white, U-shaped arch between two banks of a large cloud, recalling French's observation of "a sort of arch" between a cell on the right and a cell on the left.

We'll probably never know the truth about flight 301. One additional clue is worth mentioning: On November 8, 1963, the day before the incident, a tornado struck Refugio County about 140 miles southwest of Houston. The 150-foot-wide twister wrecked two homes and injured one person. Of course, it wasn't the same tornado that supposedly hit flight 301 (tornadoes don't last an entire day). However, the Refugio twister was the *only* officially recorded tornado in the state of Texas between May 29, 1963, and March 18, 1964. The coincidence of its timing, and its closeness to the Houston area, makes one wonder if other, unrecorded funnels brewed over southeast Texas skies at 3 P.M. on November 9, 1963.

Bates warned that airplanes might fly into tornadoes again. He noted that aerospace firms were designing the mega-passenger planes of the next decade—leviathans that would haul hundreds of passengers at a time. Aerial holocausts were possible.

Bates urged scientists to learn more about the tornado threat to aviation, perhaps by "dropping or launching . . . tracers and probes" into funnels. Others had similar ideas. Soon they would turn their ideas into action—not on tornadoes but, rather, on their watery counterpart, the waterspout.

What is a waterspout? Is it just a tornado over a water surface . . . or is there some fundamental difference in structure and energetics?
—Joseph H. Golden, *Monthly Weather Review*
(1971)

In the late 1960s, a young doctoral student from Florida State, Joseph Golden, was flying with a friend over the Florida Keys. It was strictly a pleasure flight. But Golden's life changed when, gazing out over the blue-green waters, he spotted a waterspout. "We were screaming and yelling, as excited as kids at a football game. I had a brand-new movie camera with me, a Super 8, and it was the first time I had ever used it." He filmed the waterspout from a mile or two away.

The sight changed his career. He and several colleagues would spend years studying the waterspouts in unprecedented detail. In the process Golden earned his doctorate at Florida State University and became (in one admirer's words) "the father of waterspout research." The scientists even risked their lives by flying *through* waterspouts in planes. What they learned would shed light on waterspouts' big brothers, tornadoes.

Golden told his superiors he'd like to study waterspouts. Initially they were skeptical; they doubted waterspouts were common enough for him to gather adequate data. Finally Golden managed to scrape together a $2,500 research grant. That was peanuts even by the standards of the early 1970s. But he was a bubbly, enthusiastic guy, a "struggling young researcher" with powerful skills of persuasion. He managed to cut costs by persuading a Key West hotel to charge him "ultracheap rates" and an airline to fly him for free. In return, he said, he would share his data with them (he hinted that waterspouts were a potential aviation hazard). He also talked the Navy into giving him smoke flares. "My enthusiasm, I think, rubbed off on a lot of people." Later he won support from the NRC. Some utility companies were considering whether to build floating nuclear power plants on offshore barges; the NRC wanted to know if waterspouts would threaten the reactors.

To track down waterspouts, Golden got help from a "spotter" group. This consisted of about 25 observers on the islands who kept their eyes peeled for funnels. The first year of his research, the spotters reported a startling 500 waterspouts.

"Key West is the greatest natural vortex laboratory in the world," says Golden, now an official of the National Oceanic and Atmospheric Administration (NOAA) in Washington. "Look at how many of these things occur in a small area. [The tornado scientist Howard] Bluestein estimates he and his students have to go out nine times to see one vortex, whereas seven out of ten times when I got out, I'll see a waterspout."

Ultimately Golden concluded that Keys waterspouts occurred ten times more often than previously thought. That, in turn, suggested waterspouts were a greater hazard than generally believed. True, most waterspouts are harmless. But some have wreaked havoc. A few waterspouts strike harbors, wrecking boats and buildings. A 1968 waterspout in Miami threw a 36-foot cruiser more than 200 feet through the air. A 1971 waterspout in Pensacola, Florida, caused $3 million in damages. A 1980 waterspout sank a shrimp boat in San Antonio Bay, Texas, injuring two; another crew member was never found. In 1993, a waterspout on Lake Michigan killed a windsurfer. Some of the most damaging waterspouts occur in Florida's Tampa Bay, which is densely urbanized.

On Golden's early research flights, he rented a plane and told the pilot to fly near waterspouts. As he passed the funnels, Golden threw smoke flares from the plane. The smoke trails revealed the direction of air currents around the waterspouts. The waterspouts appeared to form along boundaries between different currents of air—say, between warm sea air and cold outflow from rainy cumulus clouds. He also tossed out balloons and confetti.

Golden concluded that all waterspouts go through the same "life cycle." The first sign of a budding waterspout is a "dark spot" on the ocean surface. A dark spot is "a prominent light-colored disc on the sea surface surrounded by a dark patch." Often the dark spots appear in long lines, as if they were forming along a common boundary between air masses. When he dropped smoke flares near the dark spots, the smoke trails curled around them. That proved air was spiraling into an invisible vortex above each dark spot. "Alternating dark- and light-colored surface bands" formed around a dark spot, crowned by a cloud of sea spray. Air pressure in the vortex dropped, causing water vapor to condense into mist. Result: The funnel appeared as a stubby column protruding from the cloud base, then linked to the ocean's surface. Waterspouts dissipated (often abruptly) when a wave of cold air rushed in.

Golden and other scientists wanted to gather information from *within* the waterspout funnels. How low did air pressure fall within the funnel? Did the vortex column contain a core of sinking air? About the same time, two scientists from Purdue University, Christopher R. Church and Charles M. Ehresman, developed an instrumented probe that they towed behind a Cessna and dragged through waterspouts. The probe was a cone-shaped object, about the size of a large typewriter, that contained instruments to measure air pressure and temperature. The cone also contained smoke flares, which could be ejected on command from the plane. Golden joined Church and Ehresman on many flights. The pilot had to maneuver the plane carefully to ensure that the towed probe passed through the core of the funnel. The maneuver required the plane to execute an arcing maneuver around the funnel; in the process, the plane sometimes passed within 30 feet of the huge vortex. Golden gazed in awe out the airplane window at the funnel, which resembled an ever-changing gossamer veil, rippling with turbulent eddies. He felt as if he could reach out and touch it.

The towed probe proved to be a major headache. Every time they dunked it into the funnel, the probe hurtled downward, "like a bullwhip." Golden said. "That caused us all kinds of grief in interpreting the pressure data."

Once they arced around the funnel and Golden warned the pilot not to get too close. Golden dreaded what might happen if the plane's wing punched through the funnel wall. He had already seen what the funnel did to the towed probe. The Keys were part of the so-called "Bermuda Triangle," where, according to local legend, numerous ships and planes disappeared; he wondered if those unlucky sailors and pilots had crashed into waterspouts. Suddenly—he couldn't believe his eyes—the Cessna's wing plunged into the funnel. "I was scared to death. I knew for at least an instant we were in great peril."

To his astonishment, the plane remained stable. The wing slid back out of the funnel, and they flew away, somewhat shaken.

Afterward, Golden reflected on the incident. Their Cessna was undamaged. Therefore, he reasoned, an even sturdier aircraft might be able to fly *through* the funnel—punch right through it, wings, fuselage, passengers, and all.

Golden discussed the possibility of flying through waterspouts with Peter Sinclair, a scientist who had chased Arizona dust devils with an instrumented jeep. Sinclair "was one of my early heroes. I thought, 'If he can do it with dust devils, maybe I can do it with waterspouts.'" Sinclair was game. They prepared to fly through the waterspouts in a sturdy two-seater aircraft, an AT-6 trainer plane of World War II vintage. They covered the plane with weather instruments to measure conditions within the funnel.

Golden grew anxious as the big day neared. "I thought, 'Golden, you idiot, are you sure you should be doing this?' I didn't even tell my wife how *risky* this was." He simply told her he was planning to fly through a waterspout, as casually as if he were planning to drive to the bank. "She just sort of shook her head."

The day arrived. The AT-6 taxied and took off. They prowled the skies until they spotted a waterspout. From a distance, it resembled a thread; closer in, a smooth veil. It hung from the asphalt-black underbelly of a cumulus congestus cloud. They flew under the cloud toward the funnel. In the growing darkness, Golden sat behind the pilot with "white knuckles and gritted teeth, holding on for dear life." A fraction of a second before they penetrated the funnel, Golden saw that its surface wasn't smooth; it was a turbulent swirl of vapor. Then he felt a jolt, his head struck the canopy, and they were inside the vortex. Less than one second later they flew out the other side.

Afterward they returned to Key West. Golden went to the beach, drank a margarita, and tried to unwind.

Golden, Sinclair, and a student flew through numerous waterspout funnels that summer. They were careful to fly only through small funnels—roughly 30 to 50 feet wide—and only at high altitudes, say, 1,500 to 2,000 feet above the ocean, where waterspout winds are less intense. Nowadays he warns boaters against venturing into waterspouts, whose lower-altitude winds may exceed hurricane force.

During his many waterspout penetrations, Golden endured just one severe jolt—the type that "throws you out of your seat, then throws you *back* in your seat." Later, examining data from instruments mounted on the exterior of the AT-6, he realized why that jolt was so bad: The vortex winds were asymmetric. That is, the wind on one side of the waterspout was much stronger than the other side. Why? He didn't know. But the observation proved important later, when Golden went tornado chasing in the Midwest and co-discovered asymmetric airflow into a tornado. It was yet another example (like Fujita's suction vortices) of the structural complexity of tornadoes.

As it turned out, waterspout penetrations were fairly safe far above the water, as long as the plane was sturdy and the pilot was competent. Golden and his colleagues weren't even the first: Navy pilots had flown through waterspouts since before World War II. Rossow also flew through funnels, but buried this fact in his obscure 63-page report to NASA. In the 1970s, H. V. Senn of the University of Miami flew gliders through waterspout funnels. Most of the time, Senn said, the penetration was no big deal: He experienced a jolt and a moment of "drastically reduced visibility," then emerged on the other side, unharmed. Once he entered a funnel at a low angle; his sailplane flipped upside down and plunged 200 feet, but he quickly righted it. "I have flown dust devils over the Mojave Desert that had much more turbulence and much greater vertical velocity." But he warned adventurers who hoped to imitate him:

> The funnel clouds reported here appear to be typical of those in south Florida, but should *not* be assumed to be typical of even small midwestern [tornado] funnels. Nor would it be wise to assume that all funnel clouds may be routinely penetrated by either sailplanes or small powered aircraft without more serious consequences than were suffered in these flights.

Stirling Colgate would soon recognize the truth of those words.

The way to practice for flying through tornadoes is to fly through mountains.
—Jim Cook, tornado-chasing pilot (1968)

The space age dawned in the late 1950s and 1960s. U.S. and Soviet scientists launched robot probes to the Moon, to Venus, to the outer planets. Why couldn't we launch automated probes into tornadoes, as well? Tornado experts had speculated about such probes at least since the 1950s. Some, such as Vonnegut, wanted to do so to test their pet theories about tornadoes. Others saw the probes as the prelude to tornado control—as the vanguards of future probes that might contain explosives, chemicals, or other tornado-busting ingredients.

In 1971, F. C. Grant of NASA proposed using a research airplane to drop an instrumented balloon into a tornado. In a 1972 study for the University of Notre Dame College of Engineering, B. J. Morgan suggested that someone should outfit an armor-plated vehicle with weather instruments and drive it near a tornado. Church, Ehresman, and Golden's towed waterspout probe was the first scientific instrument "launched" into a major atmospheric vortex.

The first true tornado probe was TOTO (Totable Tornado Observatory). Its acronym was an allusion to Dorothy's dog in *The Wizard of Oz*. The 400-pound, cylidrical probe was deployed in the early 1980s by Howard Bluestein and his colleagues from the University of Oklahoma. They rushed around the countryside, chasing tornadoes, with TOTO on the back of their truck. When a twister neared, they pulled over, unloaded TOTO—the operation took 30 seconds—and scrammed. Despite their many tries, no tornado ever passed directly over TOTO. Bluestein and a colleague blamed the failure on "the short-lived nature of most tornadoes and the coarseness of the network of good roads." TOTO was decommissioned and is now stored at NOAA in Washington. It's the ancestor of the "turtles" deployed by VORTEX in 1994 and 1995.

✦

Stirling Colgate traces his fascination with tornadoes to a lifelong love of explosions. He explains that he grew up in the countryside, where "you're always involved with dynamite. On farms back in the 1930s, you didn't have tractors to uproot stumps. You had dynamite." He has spent most of his career studying different types of explosions—from nuclear weapons blasts to exploding stars to accidental explosions at nuclear power plants. "And the tornado is a kind of atmospheric explosion."

Despite his rural upbringing, he was the child of privilege—the heir to a famous name and a huge family fortune based on toothpaste. Science fascinated him as a boy. "I was called a 'professor' at home when I was four or five years old. I suppose I made my first electric motor when I was eight." He attended the prestigious Los Alamos Ranch School in the mountains of New Mexico. In the early 1940s, as America entered the Second World War, two mysterious men arrived at Los Alamos. One wore a porkpie hat and called himself Mr. Smith; the other sported a fedora and identified himself as Mr. Jones. Shortly after their arrival, the erudite young Colgate glanced through a textbook, saw their pictures, and realized their real names: J. Robert Oppenheimer and Ernest O. Lawrence. "At that moment, I knew what they were working on, even though it was secret. They were planning to build an atomic bomb."

Los Alamos was converted from a boys' school to the nation's first atomic weapons laboratory. After growing up and receiving his doctoral degree at Cornell, Colgate worked at 3 federal labs—including Los Alamos—on nuclear bombs and other subjects. It was a highly secretive world, whose intellectual elite couldn't casually chat about their work; spies might be listening. Much of their greatest research was stamped TOP SECRET. In astrophysicist Kip Thorne's book *Black Holes and Time Warps* (1994), he recalls hiking in the High Sierras with Colgate, "one of the best American experts on the flows of fluids and radiation." They talked about physics. Thorne asked Colgate about an astrophysical phenomenon that involved x-ray radiation. Did the phenomenon really behave as theory said it should? "It has been shown," Colgate replied vaguely. Puzzled, Thorne persisted: "Where can I find the calculations or experiments?" Colgate: "I don't know." Suddenly Thorne realized why Colgate couldn't explain the phenomenon to him. Bomb makers employed a similar process to trigger the explosion of a hydrogen bomb—the TOP SECRET to end all top secrets. Scientists love to blab about their work, but for many, the Cold War sealed their lips.

Colgate didn't just work on "nukes," however. The nation's nuclear weapons labs were open to people with wide, even eccentric, interests. Through his career, Colgate has ventured into subjects far from his core interests or expertise—volcanoes, AIDS epidemiology, nuclear power, cosmology, tornadoes. Some projects are unusual, to say the least. He worked for a while on a project with Charles B. Moore to eject electrically charged particles into the atmosphere in hopes of understanding how the charge on clouds might affect rainfall. After observing volcanic steam explosions, Colgate wondered if steam explosions might happen inside nuclear power plants. He wrote an unpublished paper on the topic. Later an environmental group discovered the paper and published it, although by that time Colgate had rejected his own work.

Colgate doesn't mind being wrong. Once he challenged a central concept of cosmology, one that involves exploding stars. The idea was incorrect, but he takes pride that "two Ph.D. theses were written to *prove* it was wrong."

Colgate became interested in weather in the mid-1960s, when he became president of New Mexico Institute of Mining and Technology in Socorro, New Mexico. He got to know scientists at the school including E. J. Workman, a thunderstorm expert working on lightning suppression, and Vonnegut's close colleague Moore. As a private pilot, Colgate helped Vonnegut and Moore to erect one of their grandiose experiments to alter the electrical charge on clouds. The experiment involved stretching long, elevated wires between two mountain peaks. Flying his own aircraft, Colgate draped the wire between the peaks.

In 1985, at age fifty, Colgate resigned as university president and retreated with his wife, Rosie, to a cabin in Ward, Colorado. "I just quit everything. I was worn out. I refocused my energies on things that interested me." He decided to revisit an old interest: tornadoes. In his 1967 *Science* paper, he had urged scientists to launch instrumented probes into funnels. Well, why not do it himself? He asked the National Science Foundation for money to develop a prototype rocket. The NSF agreed on one condition: that the Federal Aviation Administration approve the rocket for flight. The FAA looked at Colgate's design for the rocket and said fine, go ahead. NSF awarded Colgate $60,000.

The "small rocket tornado probe" was the length of a cheerleader's baton and weighed less than a pound. The probe contained tiny weather instruments, powered by a 9-volt alkaline transistor battery. He planned to strap several rockets to the underside of his plane's wings, then fire them into tornadoes. The instruments would measure air pressure and temperature within the funnel, plus—shades of Vonnegut—any electrical activity. The probe's radio transmitter would send its measurements back to a minicomputer aboard the aircraft. Colgate developed numerous probes with the assistance of students and various friends in high-tech industries.

But he didn't want to test the rockets on a tornado, not right away. That was too risky. Instead, he did what Rossow did—he headed to the Florida Keys, where he would test the rockets on waterspouts. He and a student lived in the Keys for two months, using Colgate's Cessna 210 to chase waterspouts. Colgate fired numerous rocket probes into the waterspouts, managing to hit "no more than a few." After a hard day of waterspout chasing, they relaxed at Key West's Half-Shell Raw Bar.

In spring 1980, Colgate decided he was ready to take on tornadoes. He shifted his base of operations to Norman, Oklahoma. Each morning he talked to meteorologists at NSSL to learn where storms were brewing. Then he flew there and surveyed the skies. From the wings of his Cessna hung a small arsenal of tornado rockets. Sometimes he'd fly day after day, never seeing a funnel. Typically "after two or three weeks, my wife, Rosie, would show up. Many times she'd stay at the airport in Norman, and stay there until midnight waiting for me to get back. It was hard on her." He chased again in the springs of 1981 and 1982.

May 22, 1981, was a busy day for tornadoes in western Oklahoma. Seven tornadoes touched down, ranging in intensity from F2 to F4. They hurt no one but caused scattered damage. The biggest twister was 1,200 yards wide, a monster that traveled 25 miles from the town of Binger to 12 miles north of Union City. It hurled cars and large oil tanks for a half mile, wrecked seven homes, and threw dead cattle into trees.

That day, Colgate encountered another twister 70 miles west of Oklahoma City. He fired four rockets at the funnel; he missed every time. The firings were recorded by an on-board Super 8 movie camera and two 16 mm movie cameras, one on each wing. The film includes Colgate's narration. At the start of the film, the tornado appeared about a mile away. It was a grayish, ragged cone, whirring across the checkered farmlands. As Colgate spoke, he seemed to be panting slightly.

"We're going slowly towards it, but we see that it's somewhat unstable," he said, "and we're going to get as close as we can safely. It is *fantastic!*" Rain streaked across the window. Colgate had already used up half his movie film, but no matter: "I don't care if that movie goes all the way to the end, it's too fantastic a sight to miss. We've got some rain there, which is screwing up photography. No sign of any turbulence." The tornado was hundreds of feet wide, sheathed in dark mist, and appeared to be breaking up. He reached for the FIRE button and pressed it. The rocket spurted from its crib on the underside of one wing and blazed ahead, leaving a contrail behind. It shot past the twister. "We missed," Colgate groaned.

Now he had to move fast. The tornado was shrinking, a bad sign for several reasons. First, a narrower funnel is harder to hit. Also, as the funnel shrinks, angular momentum makes it rotate faster. The faster the twister rotates, the greater the chance it will tear the probe to bits. Colgate fired again: "One, two, three—fire!" Another rocket shot forth; another miss.

He decided to go for broke; he would move in closer. Now he was only a few thousand feet from the funnel. "All right—three, two, one: fire! There." The rocket shot straight toward the twister. Gleeful, Colgate started to shout "Perfect!" But he didn't get past "Per—" The rocket veered away and vanished in the distance. He moaned. The day's chasing was finished, and all for naught.

The following January, Colgate described his work at the American Meteorological Society's 12th Conference on Severe Local Storms, in San Antonio. "The [rocket] equipment needs to be hardened and engineered to a significant degree," he said, "but we believe we have proved the feasibility of the probe, tactics, and launch platform [airplane] for future tornado work." He suggested firing future probes "from a land vehicle or a helicopter."

Colgate's third and final session was the spring of 1982. On May 19, he and a student, Dan Holden, soared in the skies over Pampa, Texas. Colgate flew a Centurion 210, designed to handle "near acrobatic stresses." On the ground, Rasmussen (then at Texas Tech) and other chasers pursued the Pampa tornadoes by car. About 5:30 P.M., Texas Tech's Tornado Intercept Program saw "striations" in the clouds "moving rapidly toward the wall cloud," as famed chaser Tim Marshall later wrote. Striations are grooves or channels in the clouds that indicate the direction of airflow. A few tornadoes came and went. "Then a large tornado with three sub-vortices developed

A tornadic thunderstorm crosses the path of tornado researchers.
[JIM REED]

A school was demolished in southern Tennessee during an outbreak of tornadoes in 1995. [JIM REED]

A tornado struck a mobile home but hurt no one on May 13, 1989, at Hodges, Texas. In its last stages it stretched into a 20-foot-wide vortex 3,000 feet high. [ROY BRITT]

A tornado hit Savasosa, Texas, causing major destruction. [WARREN FAIDLEY]

A tornado approaches Tucson, Arizona, in March 1994. It caused minor structural damage. [WARREN FAIDLEY]

Two of the top storm chasers in the country: VORTEX director and field commander Erik Rasmussen (top) and VORTEX assistant director Jerry Straka. [JIM REED]

Professor Charles Anderson, pioneer of tornado debris research.
[UNIVERSITY OF WISCONSIN—MADISON]

The beginning of a supercell storm in West Texas. [WARREN FAIDLEY]

a mile northwest of our location. . . . The tornado traveled northward through a plowed field picking up loose soil which changed the tornado to a dark red color. . . .

"By 7:05 P.M., the large, cyclonic tornado reversed its direction and moved southeastward toward our location. As a result, the team had to make a fast exit toward the east. . . ." The tornado hit the Pampa Industrial Park and smashed seven buildings.

Up in the air over Pampa, Colgate was having his own problems. He was chasing a twister when, suddenly, the plane hit a powerful downdraft. The aircraft reeled and dove toward the ground. He wrestled with the controls; Earth sped toward him like a wall. He turned the aircraft around moments before hitting the ground. Just before he righted the plane, he swears he saw "the television picture in someone's living room."

Shaken but determined, he flew on to the nearest tornado, "fired two or three rockets and turned tail." On the way back to Norman, a furious thunderstorm pummeled the plane. Colgate made an emergency landing in a field. He was shaken and upset. He would never chase another tornado.

Colgate feels guilty about that last chase: "I took too great a risk with the student's life." (Holden is now a scientist at Los Alamos.) Over three years Colgate had chased several twisters, come within a few hundred yards of one, and launched a total of "at least a dozen rockets." About half missed, while technical problems prevented the others from sending useful data. One problem was the rocket's fins, which sometimes sheared off in flight.

Colgate's friends and associates worried about him. They feared that he would overcome his jitters and fly again—the next time, fatally. One day he encountered a friend from the National Center for Atmospheric Research on a street in Aspen, Colorado. The friend, a leading theoretician, assured Colgate that he didn't need to fly any more missions because theoreticians had just solved the problem of tornadoes. Their new computer models satisfactorily explained how tornadoes acquire their energy, and (best of all) in terms of known physics. "Oh, Stirling," the friend cooed, "you should look at some of the calculations. We've got this whole thing nailed."

Three years later, Colgate ran into the friend again. By that time the friend was singing a different tune: They hadn't figured out tornado physics after all. Colgate reminded the friend of his former enthusiasm. The friend replied, "Oh, I didn't really know that we had solved it. I was just trying to save your life."

"And then," Colgate adds with a laugh, "I knew I had friends."

Nowadays, scientists could deploy far more sophisticated probes than Colgate's. Modern microchips store thousands of times more data than was possible in the early 1980s. Perhaps the future lies in remotely piloted vehicles, like the "spy planes" with video cameras that are being snapped up by the armies of the world. In that regard, researchers at the University of Oklahoma have been developing a remotely piloted, video-equipped plane that might fly near tornadoes.

What Colgate really lacked, he now says, was a "sizable organization. It was for too big for me, in complexity." He was discouraged and gave up after his near-fatal flight in 1982. But his dream never died. Has he ever thought of trying again? He laughs and says, "I just had my seventieth birthday, and the answer is, *yes!*"

Erik Rasmussen of VORTEX hopes that Colgate-like rockets eventually make a comeback. "Someday we'll need to get inside the tornado. Maybe we'll be driven to go there by curiosity, nothing else." He's skeptical about the handful of anecdotes told by people who claim to have looked upward into a tornado funnel and seen fantastic lightning (Will Keller) and glowing lights (Roy Hall). "If I had heard fifty anecdotal accounts over the last half century of people being inside tornadoes and looking up and seeing lightning dancing around inside the funnel, I would take that seriously. But *one* account? It's not even worth giving much thought to.

"It would be *nice* to know if lightning is dancing around inside a funnel, although I'm not sure how important that is. But I would really like to know what the temperature is like inside, and the pressure distribution." Why temperature and air pressure? Because the distribution of temperature and air pressure inside the funnel would help scientists to decide whether air is sinking in the core of the tornado; sinking air warms as it descends and is compressed by higher pressure. According to theory, suction vortices are generated by sinking air within tornadoes: As sinking air falls through the funnel and collides with rising air, the collison creates turbulence that generates the mini-vortices. In short, by mapping the temperature and pressure distribution within different tornadoes, scientists could clearly determine what causes suction vortices and, thereby, why some tornadoes are more devastating than others.

But they probably won't map temperature and air-pressure distribution within a funnel until they can thoroughly probe their interiors—at different altitudes—with instrumented probes like Colgate's. Rasmussen doubts unmanned aircraft could survive the environment near a tornado, so rockets remain the best bet.

"I actually was talking to a contractor this spring [1995] about developing a missile system for me," Rasmussen said. "He said he could do it for about $25,000. We just didn't have the money to spend on it."

Until the money shows up—and it isn't likely to show up as long as Congress slices and dices the federal science budget—the best "instrument" for studying tornadoes remains the human eye. And hundreds of eyes are available for that task every spring—the eyes of amateur tornado chasers who, despite their eccentricities and occasional recklessness, have become the irreplaceable People's Army of tornado research.

CHAPTER 6

On the Road

Tornado studies are one of the few fields of scientific research where the amateur can make a significant contribution.

—Thomas P. Grazulis

I shudder to think of what a feeding frenzy the media will have on the day when a chaser gets killed by a severe storm. . . . When it happens, there may be talk of banning the "sport" or regulating it or whatever. I hope that I don't live to see the day.

—Charles A. Doswell III

A new subculture is sprouting across the Midwest: amateur tornado chasers. Mostly young and male, they race up and down the Midwest for days or weeks at a time, searching for the nastiest storms. Many chasers live off junk food and sleep in crummy motels. As the days pass, their cars fill with used videotapes, Coke cans, and Big Mac cartons. In short, they're having the time of their lives. And like any young people having a great time, they've made enemies.

The typical chaser is like Bob Henson: intelligent, level-headed, and mature. He'd never dream of driving onto private property to "get a better angle" for his video shot, or barging into a rural National Weather Service office and using the computers without permission. But a minority of chasers do these things, and they've made life harder for those who don't.

Recently, the misfits earned a tongue-lashing from one of their heroes: Charles "Chuck" A. Doswell III of NSSL. Doswell is an aging, cantankerous cowboy type (tall, skinny, Burl Ives goatee, wears a big cowboy hat). He joined the first scientific chase teams in 1972. In March 1995, he was a star speaker at the first storm chasers' conference at a hotel in Norman, Oklahoma. Chasers and "wannabe" chasers packed the room (twice the expected number showed up). Doswell glared at the assemblage and warned: "If you choose to behave stupidly, there is nothing I can do about it. The highways are as open to you as they are to me.

"However, if you do something really dumb, and that action in some way jeopardizes my opportunities to chase storms, then I reserve the right to be upset with your behavior and to create as many obstacles to your continued stupid behavior as I can think of and get implemented."

He added, darkly: "If you ruin it for me, I'll come looking you."

The perennial fear of chasers is that law-enforcement authorities will finally crack down and "ban" chasing (although it's unclear how a ban might be enforced). A ban could be triggered by a single dreadful incident: say, a chaser who, while pursuing a funnel cloud, slams into a school bus. Fifty little charred corpses—that would kill chasing, for sure.

Several years ago a lone chaser, a student at the University of Oklahoma, skidded on a wet road, flipped over, and died. His wasn't an "official" tornado death because he was killed by a puddle, not a funnel. But human carelessness and the law of probability guarantee that one grim day, a twister will catch a chaser and wrap his car around him. Several chasers told me they are amazed that it hasn't already happened. There have been *so many* close calls. . . .

A major risk is "chaser wannabes," people "who want to see a storm but have no idea what they're doing," says Gilbert Sebenste, a noted chaser from De Kalb, Illinois. He worries about the potential for a "major disaster." "If a tornado decides to take an abrupt turn—and they can do so at any time—everyone will be scrambling to turn their cars around, and a lot of people will die. That's one of the biggest fears that all the storm chasers have now. It has gotten so crowded out there that there is no room for error."

Chasing poses another risk—that it will swallow the chaser's soul. Young people have abandoned friends, families, and everyday reality for the netherworlds of computer hackerdom, or Dungeons and Dragons, or assorted religious cults. Likewise, "chasing can become a dangerous obsession," Doswell said. "Student chasers who let their studies go to follow the convection run the risk of sacrificing their careers. Chasers who neglect their family responsibilities to chase are not folks I admire, no matter how much they 'succeed' in chasing. . . . Some folks wear their obsession about chasing as some sort of badge of honor, but I say, 'Get a life!' "

Those are ironic words coming from one of the nation's pioneering storm chasers—a man who, in 1995, risked his neck to record what may be history's most incredible close-up video of a tornado. Is Doswell just another old grouch, grumbling that the kids are going to hell? Why do they lust after a phenomenon that most sane people would rather avoid? Why, in short, do they chase?

*T*he first thing to remember is that amateur chasers serve a genuine scientific purpose. They are valuable partly because of their sheer numbers: They see and record tornadoes that might otherwise go unnoticed, and hence allow scientists to improve their understanding of tornado climatology—the frequency of tornadoes in different regions. Also, by spotting tornadoes, they enable forecasters to learn whether their forecast of a tornado came true or not. Forecasters can't improve their tornado forecasts unless they have a better idea what conditions are most likely to generate tornadoes. Amateurs' observations are partly responsible for forecasters' growing (and unsettling) realization that only a fraction of mesocyclones lead to tornadoes.

Unlike VORTEX scientists, amateur chasers aren't dependent on federal funding; they're out there on the highway every spring, hundreds of them, armed with video cameras and in some cases more sophisticated equipment. Amateur tornado chasers are part of a larger tradition of amateur science that has made significant contributions to our understanding of nature. (One of the most distinguished modern examples is the American Association of Variable Star Observers, amateur astronomers who record the changing brightness of certain stars.)

Still, judging by the gleams in their eyes, amateur chasers would chase even if their efforts were scientifically worthless. Simple curiosity is adequate motivation.

Of course, curiosity is a dangerous trait. "Curiosity killed the cat," we are warned. Youngsters who love science and nature are scorned as "nerds." Fortunately, the brightest youngsters love science and nature more than they fear ridicule; they maintain their inquisitiveness about the Earth and universe into adulthood. Otherwise, the United States wouldn't win the lion's share of Nobel prizes in science year after year.

Storm chasing is one of the bravest expressions of scientific curiosity. A person who is curious about great literature doesn't risk his or her life by opening a book. An art lover doesn't fear death upon entering the Guggenheim. But storm chasers are so enthralled by one of nature's grandest events—the thunderstorm—that they risk *death* to observe it. To hostile outsiders, this impulse appears reckless, even psychopathological.

If that were so, then Benjamin Franklin belongs in the front ranks of psychopaths. He was America's first recorded "storm chaser." In 1755 he, his son, and friends, including a Colonel Tasker, were riding horses in Maryland when

> we saw, in the vale below us, a small whirlwind beginning in the road and showing itself by the dust it raised and contained. It appeared in the form of a sugar loaf, spinning on its point, moving up the hill towards us, and enlarging as it came forward. When it passed by us, its smaller part near the ground appeared no bigger than a common barrel, but, widening upwards, it seemed at forty or fifty feet high to be twenty or thirty feet in diameter. The rest of the company stood looking after it, but, my curiosity being stronger, I followed it, riding close by its side, and observed its licking up in its progress all the dust that was under its smaller part.

Then Franklin—the same man who risked electrocution in an experiment with lightning—experimented with the vortex.

> As it is a common opinion that a shot fired through a waterspout will break it, I tried to break this little whirlwind by striking my whip frequently through it, but without any effect. Soon after, it quitted the road and took into the woods, growing every moment larger and stronger, raising instead of dust the old dry leaves with which the ground was thick covered, and making a great noise with them and the branches of the trees, bending some tall trees round in a circle swiftly and very surprisingly, though the progressive motion of the whirl was not so swift but that a man on foot might have kept pace with it; but the circular motion was amazingly rapid. By the leaves it was now filled with I could plainly perceive that the current of air they were driven by moved upwards in a spiral line; and when I saw the trunks and bodies of large trees enveloped in the passing whirl, which continued entire after it had left them, I no longer wondered that my whip had no effect on it in its smaller state. I accompanied it about three-quarters of a mile, till some limbs of dead trees, broken off by the whirl, flying about and falling near me, made me more apprehensive of danger; and then I stopped, looking at the top of it as it went

on, which was visible by means of the leaves contained in it for a very great height above the trees.

. . . Upon my asking Colonel Tasker if such whirlwinds were common in Maryland, he answered pleasantly: "No, not at all common; but we got this on purpose to treat Mr. Franklin."

Franklin's descendants now jam the highways every spring, angling for the best views of a supercell.

The first of them was David Hoadley. Now he's in his late fifties (even older than Chuck Doswell and Robert Davies-Jones). At the Norman conference in 1995, Hoadley chatted with chasers in their twenties, a few of whom probably wondered, "Who's this old dude?" Professionally, he's a budget analyst for the Environmental Protection Agency. Privately, he's the father of modern storm chasing. Almost two decades ago, he founded *Storm Track* magazine, which remains (under a new editor, Tim Marshall) required reading for chasers and wannabes.

Sometime in the 1940s, young David Hoadley sat in a movie theater in Bismarck, North Dakota. Then he felt a hand touch his shoulder. It was his dad, who said: "There's a better show outside in the street." They walked outside. The Bismarck streets were a wreck—overturned trees, roofless homes, flooded avenues. A twister had struck town during the movie. Ever since, Hoadley has been hooked on storms. He started chasing in 1957, while he was a teenager. Over the years, in different interviews, he has described chasing in quasi-mystical terms. He once said that chasing is "an experience with something infinite . . . [with] powers and scales of movement that so transcend a single man and overwhelm the senses that one feels intuitively . . . something eternal." Another time he said: "In the middle of a huge storm, with its wind and lightning and thunder, its grandeur and power, I experience a great exhilaration. I think, if this small part of nature seems so wonderful, isn't it incredible to imagine what an eternity in the presence of God would be like?"

Hoadley has put his finger on something. Whatever one's philosophical or religious views, or lack of them, it's hard to think about tornadoes in ordinary terrestrial terms. They seem to be "more" than mere physical events, reducible to fluid dynamics equations. They belong to that class of natural phenomena that, in popular culture, *symbolize* something larger than themselves.

In literature, a fox is not merely a fox; it is a metaphor for cunning. An owl represents wisdom; a dog, loyalty; and so on. Likewise, to our nineteenth-century ancestors, who clung to their Bibles as they rode mules westward, tornadoes embodied the fearsomeness and cruelty that *their* ancestors had attributed to supernatural beings. Frontier writers invested tornadoes with "demonic" qualities and compared them to "the angel of death." Of all meteorological phenomena, twisters are the easiest to anthropomorphize, to fantasize as conscious entities. An 1879 Missouri tornado stretched "into a serpentine-like form, hung up by the head and writhing in agony, its tail curling and lashing as if actuated by the impulses of a living body." An 1896 Texas twister looked like "an elephant's trunk in search of food." The St. Louis preacher quoted in chapter three depicted the 1927 tornado as God's tool for "chastising" humans.

Even in this book, this writer has referred to a tornado as a "monster." We say that a tornado "attacks" or "assaults" a city, as if it were alive. It is not alive, of course. But something about the tornado—perhaps its resemblance to the snake, the most sinister of symbols—tempts us to regard it as if it were alive. We perceive it in a way that we would never perceive an avalanche or a tsunami. (There are no "avalanche chasers" or "tsunami chasers.") Perhaps that temptation partly accounts for the twister's allure—and for the packed Midwestern highways every spring.

Guilt. Many experienced chasers feel it, from time to time.

Bob Henson is a Colorado writer in his mid-thirties. In

April 1991, he watched a tornado hit Andover, Kansas. It was a spectacular sight, one of the fiercest Midwestern tornadoes of recent years. Afterward he drove through an eerie blizzard of pink particles. It was insulation from mobile homes, torn to bits by the tornado.

That night he lay in his motel bed and wrestled with his conscience. Tornado chasing, he decided, is as bad as ambulance chasing.

Later he changed his mind. He reminded himself that tornadoes aren't wild monsters. Rather, they are intricate physical phenomena that "bring a hundred equations and theories to life."

"Even a tornado is incredibly beautiful in its own way . . . not the destruction it causes but the tornado itself. Part of the joy of chasing is you lose your ego. You're looking at this phenomenal storm and there's no room to be petty. . . . You're in a trance. . . . And there's the freshness of being out on the plains, smelling the weeds out in the country where there's nobody else around. It's like a show that is going on just for you."

He still spends ten to fifteen days a year pursuing awful weather across the Great Plains. The rest of the year he's a writer for the National Center for Atmospheric Research in Boulder, Colorado. When he tells people he chases storms, they ask: "You actually go out and try to *find* tornadoes?"

He does, although—ironically—he fears them. Once, he entered a Colorado hailstorm that dropped four-inch hail. He retreated to a restaurant bathroom "and put my face in my hands. I was really afraid." He suspects that, half-consciously, he chases storms to come to grips with his fear.

His fear is quite rational, though. A chaser must appreciate what a storm can do to his body. Otherwise he's an accident waiting to happen. Henson recalls encountering a 100 mph gust front near Akron, Colorado, in 1990. The gust front created a solid wall of dust, several hundred feet high; it was as if he had gone back in time to the Dust Bowl of the 1930s. "I was petrified. . . . The dust cloud was churning and rolling. . . .

It would be hard to outrun in my car. So I parked my car and went 20 or 30 yards into the field and lay down on the ground.

"Within 30 seconds I was pelted with rock and grit—little grit particles, hitting me at 100 mph! I was literally yelling with pain.

"When I got up I was covered with dust. It took me a day to get the dirt out of my hair."

*C*arson Eads has "the best-equipped chasemobile in Tornado Alley," according to the agenda at the Norman chasers' conference. In a slide show, he described how, while on the road, he uses a cellular phone, modem, and on-board computer to access the latest weather data. The van's TV antenna allows him to watch the Weather Channel anytime he likes. He also tows a portable Doppler radar behind the van. (The audience "oooohed" in envy.) An audience member asked Eads if he was guilty of technological "overkill." Eads grinned good-naturedly and replied, "Evidently I've got more money than I know what to do with."

Some chasers have turned chasing into cash. They sell their tornado videotapes to the Weather Channel or other outlets (one woman sold her tape to a beer company for a commercial). Others offer paid chase "tours." An ad in *Storm Track* for the "Tornado Alley Safari" says: "Come and enjoy the beauty and splendor of awesome skies, wild thunderstorms, and possible tornadoes. . . . One week/$1,050, two weeks $1,900. . . ."

Speakers offered tips on making and selling tornado videos. A frequent tip: Watch your profanity. One chaser, Chris Novy, has coined the word "stormgasm" to describe "the animal-like sex sounds that chasers typically make while watching a tornado. Anyone who has seen the more popular chase tapes can recall the 'oh baby . . . oh baby' grunting and howling sounds."

"Greed" drives some chasers, complained Rich Thompson and Roger Edwards in a recent essay for *Storm Track*. They accused certain unnamed chasers of rushing to TV stations to sell their videotapes rather than remaining in the field and collaborating with other chasers. Commerce replaces camaraderie. "If chasers continue to promote this sport in this manner, the roads will eventually be clogged with so many 'yahoos,' and we will have no one to blame but ourselves."

Storm Track editor Tim Marshall and Randy Forey defended storm-chase capitalism. Storm chasing, they observed, is costly (the gas bills can be murder). Selling videos helps to recoup the costs. Besides, tornado videos educate the public. They added caustically that "the 'yahoos' are not from outside—they are from within—many are well-known, experienced chasers."

All in all, the Norman conference "showed that spotters/ chasers can remain civilized in groups, and we do have something worthwhile to contribute to the science of meteorology," Marshall wrote. "We dispelled the attitude that we are just a bunch of pack animals joyriding around the countryside trying to upset law enforcement officials."

Most chasers became interested in weather at an early age. Henson was seven when he saw a tornado watch on an Oklahoma City TV station. Afterward he cut and bent milk cartons into an anemometer, which he installed on his roof.

When Sebenste was eight, he attended a picnic sponsored by his father's company in Illinois. The weather forecasters had forecast sunshine. The day turned hot and humid. Young Gilbert knew enough about weather to know that heat and humidity presaged storms. "All of a sudden, I looked off to the west and saw this big huge mushroom [cloud] going up. . . . I ran back to the picnic and said, 'Dad, there's going to be a big storm.' " Soon the downburst hit: "All the prizes got blown out into the field—the bingo chips, all the cards. . . . I got under the table, my head under my hands. Everybody was screaming and yelling. It was general chaos. My dad put me on his shoulders and started running for the car. . . . He got almost knee deep in mud. That's the first time I ever heard my dad swear." When they arrived home, "my dad launched into a tirade against the weathermen." Gilbert agreed and thought, "Hey, I can do better than these clowns." He started making his own forecasts. Now he's both a prominent young storm chaser and a weather forecaster for a fair-size TV station in Illinois.

Recently Sebenste created the Storm Chaser Homepage on the World Wide Web. The Web is the Rodeo Drive of the global computer network known as the Internet. Anyone with a computer and modem (preferably a fast one, at least 14,400 baud) can access countless Web "sites" such as the Storm Chaser Homepage, which, despite its name, actually contains hundreds of pages of information. Juicy contents include high-resolution color photos of tornadoes, technical articles, and tornado experts' essays about tornado chasing and research. One can learn much about tornadoes by pointing and clicking through the Storm Chaser Homepage—and there is a great deal to learn. Storm chasing is not for the ignorant.

One of the first things a young chaser learns is that storms have their own "geography." Some parts of the storm are much more likely to generate tornadoes than others. The reasons for this are complicated, but are related to the internal airflow in a supercell as it moves from southwest to northeast (the usual path of U.S. storms). Experienced chasers head for the storm's rain-free southwest quadrant, where (in the Midwestern United States) a tornado is most likely to form. They keep their eyes peeled for a wall cloud—the rotating cloud (sometimes shaped like a hockey puck or a wedge) that drops from the base of the thundercloud. When they see a wall cloud, they try to situate themselves southeast of it, partly to watch it approach from the west. (That lengthens their viewing time, just as one situates oneself ahead of a parade to enjoy it for a longer time.) A southeastern position is also good because, there, a chaser's view is less likely to be obscured by rain and hail, which usually fall on the northern side of the storm. And there's a third reason for staying to the southeast: to get the best photos and video coverage. Because most tornadoes occur in the late afternoon or early evening, a southeastern position is ideal because the clouds will contrast dramatically with the bright western sky, where the sun sets.

This writer visited Sebenste at Northern Illinois University in De Kalb. There Sebenste regularly updates the Storm Chaser Homepage at a computer workstation in the meteorology department. He's bubbly and extremely friendly, a tall, slender young man with short dark hair and dark-rimmed glasses. On the wall is a photo of a previous TV weather forecaster who made it big, David Letterman. Sebenste switched on the computer and signed on to the Storm Chaser Homepage. He clicked the mouse to show a page about one of his own recent tornado chases.

"It's been a very busy year. We've had a record number of tornadoes in Illinois—58 so far. It's the best year we've had since 1990 or '91. 1991, '92, '93, and '94 were just *awful*. . . . Every year is different and nobody knows why." Personally, he suspects the jet stream, that meandering river of high-altitude, high-speed air. Sometimes the jet stream diverts severe weather north or south of Great Lakes states such as Illinois, depriving him of the storms he loves.

Chasing is fun "for about three minutes out of the year. Storm chasing requires an incredible amount of patience and understanding. It can best be described as 'hurry up and wait.' You go to where the tornadoes are supposed to be and wait—hours and hours and hours. You'll be sitting under the blue sky, getting a nice tan, and that's all you'll see for the rest of day." But when a tornado touches down, "it's incredible fun and excitement. . . . You get an appreciation for what goes on in Mother Nature—appreciation that you just can't get in the classroom."

Just down the road from Sebenste is the College of Du Page in Du Page, Illinois. Professor Paul Sirvatka teaches an undergraduate class on storm chasing. Every spring he takes several dozen students on chases across the Midwest. The class helps his students visualize weather phenomena in "three dimensions," which is easier to appreciate in the field than in the classroom, he explains. He has taken students within a mile of a tornado. "We've been in some fairly hairy situations—we really thought the vans were about to tip over—but nothing I would consider life-threatening.

"This year was one of the worst—two trips, two busts. We spent most of the time sightseeing in Colorado and South Dakota, and there was just *nothing* happening in the sky. Then we came back, and while driving through Illinois, we got three tornadoes just outside of Springfield!"

Kim Ball joined that trip. She's a mother, thin and blond; she recently returned to school to get her degree in computer science. She says one of the Illinois tornadoes resembled a huge, V-shaped wedge. "Our eyes were popping out. We saw trucks and cars that had flipped over."

When this writer told her she was the first female chaser he had met, she laughed. "I've been on chases where there were three or four women, and sometimes it was almost 50-50 male and female. But when it comes to meteorology majors, there're definitely more men than women."

Anton Seimon has only a bachelor's degree in geography, not meteorology. But he's published his research on a controversial area of tornado science in the *Bulletin of the American Meteorological Society*. (Two top NSSL scientists took his work seriously enough to publish a lengthy, critical response.) He was born in South Africa in 1965. At age three, he and his family encountered a twister while driving through eastern South Africa. "The sky was black as midnight. . . . My dad told us to spread out because he thought that would prevent the car from tipping over." Later the "very left-leaning" family moved to America to escape South Africa's right-wing, pro-apartheid government of the time. The Seimons settled in the suburbs of New York City. "All of a sudden I found myself in a country with squall lines and blizzards. I was in heaven!"

In high school, teachers reprimanded Anton for "plotting weather maps instead of taking notes." He attended the State University of New York in Albany, where he befriended Bernard Vonnegut. Now in his early eighties, Vonnegut still suspects that atmospheric electricity helps to drive tornadoes. Seimon became interested in the school's lightning detection network, an array of sensors that can detect lightning strokes across the nation. He did research on the Plainfield, Illinois, twister of 1990. He concluded it was accompanied by a highly unusual amount of "positive" cloud-to-ground (CG) lightning (which transfers a net positive charge to Earth, rather than negative as usual). His *Bulletin* paper argued that the lightning data suggested "a relationship between this tornadic thunderstorm's dynamics and electrical activity."

Chuck Doswell and Harold E. Brooks of NSSL wrote a lengthy and skeptical reply. They called Seimon's paper "very interesting." But they cautioned him that most severe tornadic storms *don't* generate unusual amounts of positive CG lightning.

Challenged, Seimon decided to find better evidence for his theory. In June 1994, he and friends drove west to chase storms. He had an exciting time, spotting "some unbelievable lightning flashes, a great distance downstream from the storm. . . . I was in total awe. It's an exhilarating feeling. But I'm also aware that if I'm foolish, I could end up dead. The biggest danger in storm chasing is that you always want to have your head out the window when you should be watching the road."

Seimon's earliest interests were astronomy and archeology. These subjects have "inherent frustrations," though. In astronomy, one perceives only a tiny corner of the universe. In archeology, the past is past, as distant in time as most of the universe in space.

In contrast, meteorology is an *accessible* science. "I can fill my car with gas and drive out to find a supercell thunderstorm."

*A*nother Sirvatka student, Jeremy Hylka, is an intense young man who gave me a guided tour of Plainfield, a Chicago suburb. On August 28, 1990, a devastating tornado hit Plainfield, killing 28 people. Ted Fujita said the Plainfield tornado was the worst he had ever investigated.

Plainfield is also an omen of what may become common in the next century: the suburban tornado. As Americans increasingly flee inner cities for the peace and quiet of the outskirts, suburbs expand like algae on a rock; and as they expand, they become ever-juicier targets for tornadoes.

In the gathering darkness, Hylka and I drove through Plainfield—now largely rebuilt—and he recalled that ghastly day. "I was living in Joliet at the time. . . . Tennis-ball-sized hail was coming down; the winds were 80 or 90 mph." Afterward he rode his bicycle around the neighborhood. "It was complete devastation."

A paramedic from the Joliet Fire Department watched in horror as people were blown hundreds of yards into a cornfield. "My first thought was that Iraq had attacked," another witness said.

Many more might have died. The two-story high school can hold 1,200 students. Classes were supposed to start the next day. What might have happened if the tornado had hit the school at the same time—3:35 P.M.—but on August 29, not August 28? The body count might have been unimaginable.

Hylka studies tornadoes partly because he hopes to improve twister forecasts and to save lives. But "partly" is the operative word; he's also drawn, like most chasers, to the sheer magnificence of the phenomenon itself, however horrific its consequences. "People like us love to see Mother Nature's fury. People think we're crazy and nuts. But we *love* these things."

CHAPTER 7

The White Plague and Global Warming

As they ran from Israel . . . the Eternal rained huge hailstones from heaven on them . . . more died by the hailstones than at the hands of Israel by the sword.

—Joshua 10:11

In early 1995, on three separate occasions, the "white plague" struck the Dallas-Fort Worth area of Texas. The white plague is farmers' nickname for hail. On March 25, April 29, and May 5, more than a billion dollars in damage occurred as hailstones plunged from the sky. The hail measured up to five inches wide and wrecked cars and aircraft on the ground. On the evening of May 5, about 10,000 people attended an arts and crafts fair in Fort Worth. Baseball-size hail began falling. "Parents cowered over their children as fist-size stones propelled at 80 mph hit them in the backs, necks, heads, and arms, causing three-inch welts," storm chaser Tim Marshall reported. Car windows exploded, spraying glass on the occupants. Incredibly, no one died.

Hailstones are dramatic evidence of the strength of thunderstorm updrafts. Blowing at hundreds of miles per hour, these winds shoot water droplets several miles high, where they freeze and fall back toward Earth. On the descent, the ice pellets gather more water, then are blown upward again, where they refreeze, fall again, rise again, over and over. Gradually they acquire onion-like layers of ice—sometimes dozens of layers. Eventually they grow heavy enough to plunge all the way to the ground. Some hailstorms smash barns and wreck cars and tractors; some slaughter cattle; some kill people.

In 1936 in South Africa, jagged hail buried part of the Northern Transvaal under three feet of ice and massacred 219 souls. In 1986, a hailstorm in China killed 100 people and injured 9,000. But hailstorm deaths have been extremely rare in the United States. In 1930 near Lubbock, Texas, hail beat a farmer to death. In 1979, hail killed a baby in Fort Collins, Colorado. Animals are more vulnerable: Farmers and naturalists have observed hailstorms that killed hundreds of sheep and tens of thousands of game birds at a time.

The U.S. record for a hailstone was one that fell at Coffeyville, Kansas, in 1970. Roughly the size of a grapefruit, it measured 5.5 inches wide and weighed 1.67 pounds.

For pioneer farmers, the psychological impact of a devastating hailstorm was unimaginable. A massive hailstorm hit the Dakotas in 1883. It "seemed that half the settlers in Spink County were pulling out—some to return to their native states, others to try their luck in the logging camps of Wisconsin and Michigan," Everett Dick records in *Tales of the Frontier* (1963).

In the 1940s, the scientist Albert K. Showalter suggested that hail triggered tornadoes. Others have proposed similar theories. In chapter four, I mentioned Colonel Rollin H. Mayer and Fritz O. Rossmann's plan to destroy tornadoes with guided missiles. They based the plan on Rossmann's theory that tornadoes were spawned by cold downdrafts generated by hailstones. As hail falls, Rossmann said, it cools the air and drags it to Earth. Critics cited an obvious flaw in the theory: Tornadoes tend to form in the part of the storm where there isn't hail or rain!

Resistance seems useless. A century ago, hucksters promoted anti-hailstorm schemes. "Hail shooting with cannon, handguns, and rockets was common in the Beaujolais wine districts [of Europe] almost to the twentieth century," says meteorology historian Clark C. Spence. "In France and Italy as many as ten thousand shots were sometimes fired at a single storm, and special mortars fitted with long sheet-iron funnels hurled smoke rings at ominous clouds." In the 1960s, Soviet scientists claimed they had used cloud seeding against hailstorms with remarkable success.

More credible research in the United States has yielded mixed results. On the one hand, in the 1970s, the National Center for Atmospheric Research in Boulder, Colorado, ran the National Hail Research Experiment; researchers seeded clouds for three years. They found *no* statistically significant evidence that cloud seeding affects hailstorm incidence. In fact, seeding sometimes worsens hail.

On the other hand, hail suppression projects are still under way around the world, cosponsored by the World Meteorological Organization. Major U.S. projects are still under way in Kansas and North Dakota. In western Kansas, managers of a multicounty seeding project claim they cut hail incidence by 27 percent over 15 years.

Recent research on hail hints that we'll see more tornadoes in the future. It's just a hint—but a tantalizing one.

Scientists are increasingly worried about "global warming," a.k.a. the "greenhouse effect." The burning of fossil fuels (from cars, industries, etc.) injects carbon dioxide gas into the atmosphere. The gas traps infrared heat (as a greenhouse traps solar heat) generated by solar radiation. Many scientists think this heat-trapping raises the planetary temperature.

The planet's average temperature has risen for about a century. Global warming may partly explain why mountain gla-

ciers are shrinking. A century ago the naturalist John Muir
sketched glaciers in the mountains of Yosemite. Today those
glaciers are considerably smaller because of local (and perhaps global) warming. Another possible sign of global warming is recent satellite observations, which indicate the global
sea level is rising. Conceivably, global warming could elevate
sea level by melting polar ice.

Would global warming worsen severe storms such as hurricanes, thunderstorms, hailstorms, and tornadoes? Scientists
are starting to study this exciting question. A fascinating
study appeared in the May 15, 1995, issue of *Geophysical Research Letters*. The author, Jean Dessens, is a French meteorologist and tornado expert. Dessens assessed the possibility
that global warming would worsen severe storms in France.
Tornadoes are rare in France: hailstorms are its worst storms.
He checked to see if the incidence of hailstorms in France has
increased since World War II. He compared hail damage from
1946 to 1992 in relation to the nation's average annual mini-
mum temperature. He found an impressive correlation between the temperature and hailstorm severity: Hail damages
rise an average of 40 percent for every 1 degree Centigrade
increase in temperature. He urges other scientists around the
world to check weather records for similar links between temperature changes and severe weather such as tornadoes, flash
floods, and lightning.

The possibility of major climate change worries the world
insurance industry. If climate change occurs and causes a
wave of devastating storms, from tornadoes to typhoons, then
the insurance industry will have to pay for much of the damage. A scientific study funded by the German insurance firm
Munich Re concludes: "A warmer atmosphere and warmer
seas result in greater exchange of energy and add momentum
to the vertical exchange processes so crucial to the development of tropical cyclones, tornadoes, hurricanes, and hailstorms." The report warns that "the imminent change in our
climate makes speedy, radical countermeasures unavoid-
able."

Hurricane Andrew, the 1992 mega-hurricane that bankrupted a half-dozen insurance companies, awakened insurors to their potential liabilities. So did massive losses sustained by global insurance firms since 1987. At a London meeting in 1993, insurors said that since 1987, they had suffered blows from 15 major disasters; 85 percent of the financial losses resulted from "windstorms" of one type or another. As a result, insurors have begun funding research on global climate change. They are also prodding state and national governments to better prepare for natural disasters—for example, by requiring wind-resistant building standards.

Hurricane Andrew "was a wake-up call," Frank Nutter, president of the Reinsurance Association of America, said at a Berlin climate conference in early 1995. "Mother Nature has gotten our attention in a way never before experienced. . . . The worry now is mega-catastrophe."

Research by meteorology Professor Kerry A. Emanuel at MIT suggests that warming would breed more hurricanes. That's worth noting because many hurricanes breed tornadoes. The record for hurricane-bred twisters is held by Hurricane Beulah, which struck Texas in September 1967 and may have caused as many as 141 tornadoes. (The actual number is in dispute and may be as low as 47.) In the 1980s, 13 hurricanes spawned a total of 162 tornadoes. The worst was Hurricane Danny in August 1985, which unleashed 39 tornadoes on Texas.

Even if global warming doesn't get out of hand, severe tornadoes may become more frequent in the next few decades. Grazulis, the tornado climatologist who advised the Nuclear Regulatory Commission, says that in the 1980s, the annual number of "violent" tornadoes averaged six per year. That was remarkably low—only half the rate of each of the three prior decades (which averaged 12 violent tornadoes per year), and even less than the 1940s (15 per year). He suspects the shortage of violent tornadoes in the 1980s at least partly accounts for the remarkable fall in the tornado death rate since the 1970s. Eventually the number of severe tornadoes may rise again. "The 1980s," he says, "were clearly a low period in violent tornado activity. . . . The possibility of an increase in violent activity, and a corresponding increase in deaths, over the next decade or two is a real one."

Why would the number of violent tornadoes vary from decade to decade? No one knows. Perhaps it has something to do with short-term climate changes, such as "El Niño," a quasi-cyclical warming of Pacific waters that affects global climate. Or perhaps it's just the result of natural atmospheric variance, as unpredictable as a gambler's lucky streak.

CHAPTER 8

Putting Tornadoes to Work

an we put tornadoes to "use"? Over the decades, a number of scholars have speculated about creating or simulating tornadoes for a variety of purposes, ranging from military applications to energy production.

Humans noticed decades (perhaps centuries) ago that they could create "artificial" tornadoes. Spinning columns of air form over large bonfires, as ancient peoples may well have observed. They may also have seen funnels associated with forest fires or volcanic eruptions. Similar phenomena are seen today: Few sights are more terrifying to a firefighter in a forest than a "fire whirlwind."

Sir Napier Shaw, a famed early-twentieth-century British meteorologist, recalled that during World War I, the military discussed the possibility of generating artificial tornadoes for use against the enemy! He said one researcher thought (wrongly) that by firing artillery shells rapidly, one after another, troops could spawn a vortex. The anonymous researcher didn't say how it might be steered toward the enemy.

Later, tornadoes did form during wartime, but only as an accidental consequence of the firebombings of cities such as Hamburg, Tokyo, and Dresden, and the atomic bombings of Japan. Thousands were incinerated by the hellish vortices.

Scientists have generated smaller "fire vortices" for research purposes. The world's ultimate "tornado machine" is outdoors and located in the Pyrenees Mountains of southwestern France. John Snow has worked with scientists there to generate artificial tornadoes from giant oil fires. "Meteotron" has 105 fuel oil burners that create the equivalent of a billion watts of energy—a gigawatt, as much as a large nuclear power plant. The father-son team of Henri and Jean Dessens began working there in the 1960s, using the facility to generate artificial cumulus clouds and tornado-like vortices. Some of the clouds grew into cumulus congestus clouds, the embryos of cumulonimbus. Snow, Christopher R. Church of Purdue, and Jean Dessens studied the vortices closely with cameras and instruments. It was dangerous work. They threw smoke grenades to see which way the tornadic winds and surrounding breezes were blowing. Some of the winds were so strong that they blew out the oil-fire burners. In one case, after the oil-fire burners were turned off, a vortex more than 100 feet wide continued rotating for about 15 minutes. Some vortices extended hundreds of feet into the sky.

In the 1970s, during the global energy crisis, some experts speculated about novel energy schemes that would tap power from thunderstorms or tornado-like vortices. W. George N. Slinn of Battelle Pacific Northwest Laboratories suggested exploiting "cloud power" by funneling waste heat from nuclear power plants into the air. This would, he said, form thunderstorms that, in turn, would drop rain that could feed hydroelectric plants. Another scheme was proposed by James Yen of Grumman Aerospace Corporation, who advocated building enclosed wind generators that would suck in outside air and, by exploiting the law of angular momentum, create a narrow, low-pressure, rotating column of air—a vortex, akin to a tornado. The wind could then be used to drive a dynamo and produce electricity.

Perhaps the most astonishing human-created tornadoes— outside of warfare—resulted from an industrial catastrophe in 1926. San Luis Obispo is a charming college town nestled in the mountains of central California, not far from the Pacific. At 7:35 A.M., April 7, a bolt of lightning struck oil reservoirs at a tank farm operated by Union Oil Company, 2.5 miles south of town. The tanks detonated. The shock wave smashed windows all over town; people flooded into the streets, thinking an earthquake had hit. The burning tanks "threw out an immense quantity of hot, burning oil which spread with remarkable rapidity over an area . . . [of] about 900 acres," a witness later wrote. "The flames leaped seemingly a thousand feet in the air. . . . At the same time violent whirlwinds began to form over the fire." Apparently a northwest wind from a passing storm caused wind shear, forcing the rising, scalding air to rotate. The faster it rotated, the more it tightened into a neat, whirling column (angular momentum at work). The human-created twister "left the vicinity . . . and traveling east-northeast about 1,000 yards picked up the Seeber cottage, just outside the tank farm, lifting it several feet in the air and carrying it about 150 feet north, where it was dropped in a field, a total wreck. Mr. A. H. Seeber and his son, who were in the house, were killed," wrote J. E. Hissong of the U.S. Weather Bureau office in San Luis Obispo in the April 1926 *Monthly Weather Review.*

"A daughter and a friend, who had just stepped out of the house, were carried some distance along the ground, but were not seriously injured . . ." Hissong continued. "A few minutes later, a whirl[wind], possibly the same one, tore the roof from the house of Mr. Banks, about one-fourth mile northeast of the Seeber home; it also demolished his garage and other outhouses, uprooted fruit trees, and drove a 2-by-4-inch beam, 16 feet long, through the boarded side of the pump house." An oil company employee saw a funnel leave the inferno: "Near the mushroom-shaped top of the funnel, about 200 feet in the air, a small shed or chicken coop was floating around in a counterclockwise direction." Later spectators reported seeing a house raised 300 feet in a different funnel: "It was carried about 100 feet northwest and dropped, completely demolished."

For five days, the region resembled the setting of an H. P. Lovecraft story, a place terrorized by hundreds of these artificial tornadoes. Subsequent study of melted iron indicated the fire may have burned as hot as 2,500 degrees Fahrenheit, roughly one-fourth the temperature of the surface of the sun. Some of the twisters rotated clockwise, some counterclockwise. "Many of them had all the characteristics of true tornadoes," Hissong assured readers of the magazine, which also ran photos of the funnels swirling from the flames. "The gyrating, writhing, funnel-shaped clouds were viewed by thousands, the white, condensing vapor in the vortices making them plainly visible against the background of black smoke. Some of the funnels appeared to be not more than one foot in diameter at the smallest part. . . . It is believed that none of the tornadoes traveled more than 3 miles from the fire zone; some debris was found at that distance."

The San Luis Obispo disaster was a preview of coming events. In World War II, bombers dropped both incendiary bombs and atomic weapons that turned their targets into forests of deadly hot vortices. In his book *Hiroshima* (1946), John Hersey describes one vortex that ripped through the remains of the city after the U.S. bomber Enola Gay hit it with an atomic bomb, nicknamed Little Boy, with an explosive force equal to about 20,000 tons of TNT. Two priests, one of whom had been injured, witnessed a whirlwind tear through a park, where it knocked down trees and scattered fragments of homes everywhere. "Father Kleinsorge put a piece of cloth over Father Schiffer's eyes," Hersey records, "so that the feeble man would not think he was going crazy."

How ironic: Scientists can *start* tornadoes—as these grisly examples reveal—yet they still don't know how to stop them. Nor do they know how to forecast them (with much reliability, anyway), or even how to penetrate them reliably with scientific instruments.

Still, the future beckons. . . . Researchers are working to overcome these and other challenges, via futuristic technologies—virtual reality, supercomputers, space satellites—that, if Father Schiffer lived to see them, might also have made him "think he was going crazy."

CHAPTER 9

No Man's Land

Where will the future take tornado science? Can tornado forecasts be improved? How will twenty-first century scientists simulate, chase, and explore these storms? Will the old dream of tornado modification be revived?

And in the meantime, what can you—as an individual and a citizen—do to protect yourself, your family, and your community against these cruelest of tempests?

This writer slid the "virtual reality" helmet over his head. Through its spectacles he saw, in the distance, a bright, boiling thundercloud. He pressed a button on a baton-shaped control stick in his hand. Suddenly, he shot into the interior of the storm. Within the immense cloud, he floated from its broad, flat base to its anvil-shaped top. He saw updrafts, represented by swirls of glowing green dots. The dots zipped past him like rocket-powered fireflies. Beneath him, downdrafts flushed from the storm's bottom and swept across the countryside. He could trigger a new updraft or downdraft just by pressing the button. He felt like the ancient god Zeus, lord of lightning: He could tell a thundercloud what to do.

A male voice interrupted his reverie: "Does the helmet fit okay?" asked Professor Bob Wilhelmson. Then the writer remembered that he wasn't soaring through a storm 10,000 feet above Earth. Rather, he was standing in a virtual reality "cage" at the University of Illinois.

The University is a scientific oasis in Urbana-Champaign, amid the corn country of downstate Illinois, the eastern edge of Tornado Alley. Here, Wilhelmson and other scientists are using supercomputers to simulate tornadoes and other severe weather. Despite its geographical isolation, the campus has always been on the cutting edge of computer science. (According to the science-fiction film _2001,_ it's where the villainous computer HAL 9000 was born.) Campus scientists use supercomputers to simulate—in gorgeous color—phenomena such as exploding stars, collapsing black holes, the flow of subterranean oil, and the wiggle of electrons moving within enzymes. Almost two decades ago, Wilhelmson used an earlier supercomputer—pokey by today's standards—to simulate a thunderstorm. Now he's refining the model and adding a thunderstorm's meanest weapon: a tornado. The supercomputer crunches night and day. Its circuits are cooled by a built-in air conditioner, lest they overheat.

> Perhaps some day in the dim future it will be possible to advance the computations faster than the weather advances and at a cost less than the saving to mankind due to the information gained. But that is a dream.
> —Lewis Fry Richardson (1922)

It was "a dream" in 1922, but it isn't anymore. Computers are now essential tools in meteorology. They're vital both for daily weather forecasts and sophisticated simulations such as Wilhelmson's. Researchers at NSSL and elsewhere have developed "algorithms"—computer programs—that scan Doppler radar screens. The algorithms check for warning signs of severe storms such as tornadoes, hailstorms, and high winds. For example, scientists are testing a new algorithm that scans Doppler radar screens and looks for mesocyclones. The mesocyclones are represented by adjacent regions of different colors, such as red and green, on the screen. Green regions mark winds blowing toward the radar, the red winds blowing away. The algorithm is so good that it heralds a "paradigm shift" in such technology, says a recent paper by Greg Stumpf, Caren Marzban, and Erik Rasmussen. In a series of tests, the algorithm correctly identified 38 out of 49 (78 percent) mesocyclones on radar screens. Other algorithms monitor the screens for hail, high winds, and other severe weather. Such algorithms could save future forecasters the task of constantly monitoring radar screens.

Could future computers *replace* human forecasters? Will the next century's tornado watches and warnings be issued by a sleepless, unsalaried "brain in a box" that monitors Doppler radar screens, satellite images, and other weather data night and day, unaided by *Homo sapiens?*

Humans have feared losing their jobs to machines since the dawn of the Industrial Revolution. Machines have already replaced humans in most jobs based on raw strength and many that depend on physical dexterity. Will twenty-first-century computers eliminate most jobs based on brainpower as well? Meteorology is a classic "brainpower" career. Harold Brooks of NSSL recently wrote a provocative essay, ominously titled "The Possible Future Role of Humans in Weather Forecasting." He fears that federal budget cuts will encourage NWS to replace many salaried forecasters with unpaid computers. "There is a significant probability that humans will be out of the forecasting loop, except for the presentation end, within 20 years." NWS management may seek to justify eliminating humans, "for budgetary or other reasons," in the belief that humans add "no value" to forecasts. Why retain a human employee if a computer can do his job more cheaply?

Brooks disagrees. Forecasters, he says, are essential for

forecasting "extremely difficult" phenomena such as convection (which spawns thunderstorms and tornadoes). "Doing it well requires a knowledge of a vast range of scales of motion and behavior in the atmosphere. Human skills of pattern recognition and information processing should continue to be critical."

All-computerized forecasts may be a fantasy, anyway. For a half century, futurists have prophesied the coming of "thinking" computers, a.k.a. "artificial intelligence" or AI. In 1963, the science-fiction writer Arthur C. Clarke (creator of HAL 9000) forecast that AI would be achieved before the year 2000. By the 1980s, though, the term AI had fallen out of favor and Clarke joked that scientists had created only "artificial stupidity." The last gasp of traditional AI research may have been Japan's much-ballyhooed "Fifth Generation" project, which led nowhere. Of course, computers do incredible things. They even mimic narrow aspects of human intelligence and decision-making (via "expert systems" and "neural networks"). But computers are unlikely to replace humans in scientific fields—such as meteorology—where fundamental questions (such as, what causes tornadoes?) are unanswered. You can't program a computer to do something if you don't know how to do it yourself.

Our understanding of tornadoes and tornadic storms has progressed greatly. . . . But we still have not established an accepted theory of how tornadoes form, and uncertainty still exists concerning extreme wind speeds and pressure drops in tornadoes.

—Erik N. Rasmussen, Jerry M. Straka, Robert Davies-Jones, Charles A. Doswell III, Frederick H. Carr, Michael D. Eilts, and Donald R. MacGorman, *Bulletin of the American Meteorological Society* (June 1994)

Depending on radar alone [to forecast tornadoes] is not likely to be an effective strategy.

—Charles A. Doswell III of NSSL (1995)

It's still awfully hard to point to a cumulus cloud and say, "In three hours this is going to be a storm producing a tornado."

—Bob Henson, storm chaser

For every question we answer about severe storms and tornadoes, we probably raise 10 to 100 more. In summing up tornado research in the last 100 years, it can best be summarized as, "The more we know, the more we realize we don't know." . . . We knew we had a long road ahead, and it's perhaps an even longer road than we thought.

—Gilbert Sebenste

The more we learn about tornadoes, the more complicated they look. Decades ago, scientists assumed there was one basic type of tornado. But chase teams have discovered beyond doubt that there are at least several types. Some descend from supercell thunderstorms with mesocyclones; some don't. Supercells generate the most violent twisters, but other atmospheric events produce "landspouts," "gustnadoes," and "mountainadoes," which can also cause significant harm.

That complexity could undermine high hopes for the nation's newest defense against tornadoes: NEXRAD. This year (1996), the nation is scheduled to complete the NEXT Generation RADar network. NEXRAD is a network of 160 Doppler radar units (also called WSR-88Ds). They measure the speed of winds, based on frequency shifts in radar reflections from precipitation particles blown by those winds. When completed, NEXRAD radar will replace old-fashioned conventional radar, some of which date from the 1950s (and still use vacuum tubes!) Meteorologists say NEXRAD radar will greatly enhance their ability to detect atmospheric circulation associated with tornadoes. The radar will also detect gust fronts, strong updrafts that may generate severe hail, and deadly downbursts and microbursts that threaten both aircraft and surface structures.

Our knowledge about tornadoes and tornado-like vortices has improved markedly since meteorologists began drawing blueprints for the NEXRAD network decades ago. Back then,

many forecasters hoped that Doppler radar would revolution-
ize tornado forecasting. True, it has made a big difference:
Doppler radar has improved tornado watches and warnings
and deserves partial credit for the sharp decline in tornado
death rates since the 1970s.

But Doppler radar may have less impact in the future.
There are several reasons why.

Initially, radar meteorologists thought that about half of
mesocyclones generated tornadoes. Doppler is ideal for de-
tecting mesocyclones. Scientists were particularly excited
when they discovered the so-called "tornadic vortex signature" (TVS) on Doppler radar. A TVS represents a small region
of fast rotation within the mesocyclone. The mesocyclone
may be a few miles across; a TVS is less than a mile wide. A
TVS may form up to a half hour before a tornado drops to
Earth. So initially, scientists hoped they could issue earlier
tornado watches by looking for TVS images on Doppler radar
screens.

But now, radar meteorologists have discovered that as few
as 10 to 30 percent of mesocyclones may lead to twisters.
Many tornadoes are formed by atmospheric processes closer
to the ground, below and possibly unrelated to mesocyclones.
Furthermore, tornadoes don't always form from the biggest
storms, or those with the strongest mesocyclones. Some mesocyclones may only create an atmospheric environment in
which tornadoes are *possible;* a lower-level event (say, low-
altitude winds) may trigger the tornado itself. (Likewise, a
pool of oil makes an oil fire possible. But you need a match to
start the fire.)

If many or most tornadoes form at low altitudes, then
NEXRAD may be less effective than originally hoped, for a
simple physical reason: the Earth is round. You may have
watched boats "disappear" over the horizon, that is, over the
curve of the planetary surface. That same curvature limits the
range of a NEXRAD radar: It can't see objects (such as thunderstorms) below the horizon. That means it can't scan the
lower sections of a distant thunderstorm. Doswell says that if

a storm is more than a few tens of miles from the radar, Earth's curvature will "virtually preclude" detection of tornadic radar signals from low altitudes in the storm.

Because a minority of mesocyclones lead to tornadoes, forecasters who issue warnings every time their Doppler radar detects a mesocyclone will issue mostly false alarms. False alarms are bad because they feed what one might call the "boy cries wolf" effect: The more false alarms, the less likely the public will trust tornado watches. Such distrust could lead to countless deaths and injuries.

As Straka notes, "During VORTEX, it was amazing how many storms we were on that had tornado warnings for hours that never produced a tornado, or maybe produced one that lasted for three seconds and did nothing [else]. As a scientist I've been very disturbed by this. This needs close attention." Straka thinks that only 10 or 15 percent of mesocyclones lead to tornadoes. (Doswell recently wrote that "the figure eventually may go as low as 20 to 30 percent.")

"I'm a meteorologist," Straka continues, "and if I hear a tornado warning, I ask, 'Did they use radar? Or did someone *see* the tornado?' "

How complicated does tornadogenesis get? Doswell's memo, published on the World Wide Web, is a parade of pessimism. VORTEX observations show that tornadoes often form from interactions between supercells and other processes "about which little is known," such as lines of clouds visible from satellites, he says. "If it is indeed the case that many, if not most, tornadoes arise through *unique* processes, then forecasting and warning for tornadoes would be much more difficult than heretofore imagined. That is, it would be difficult to generalize about tornadogenesis because each situation would be unique."

In short, Doppler radar won't solve tornado forecasters' problems. For these reasons, Doswell warns against putting too much faith in fancy new computer programs that detect tornadoes in radar data. Forecasters should continue to use other ways to forecast tornadoes. For example, forecasters can monitor breaks in power lines that may be caused by twisters. They can also stay in contact with spotter groups, whose thousands of volunteers watch the skies for twisters during tornado season.

In other words: Radar is a great tool, but the best tornado detection device is still the human eye. Computers are great, but they won't replace human forecasters for a long time.

Still, hope springs eternal. . . . Better tornado forecasts may eventually come from above—from outer space.

A new space satellite monitors lightning from space. The satellite may help scientists answer an old question: Do tornadic storms generate unusual amounts or types of lightning? And if so, could they use lightning detectors on the ground or in space to improve tornado warnings?

In chapter 4, this writer talked about Bernard Vonnegut's idea that electricity generates tornadoes. That idea has largely fallen by the wayside. Still, some scientists suspect that tornadic storms tend to generate remarkably intense lightning (especially positive cloud-to-ground bolts) for reasons unrelated to the tornado itself. In the early 1950s, Professor Herbert L. Jones of Oklahoma A & M College claimed that tornadic storms emit much more high-frequency electrical signals (called "sferics," after "atmospherics") than ordinary tornadic storms. Inspired, entrepreneurs marketed "tornado detectors"—alarms that sounded if they detected a surge in the local electrical field. In the 1970s, scientists investigated claims that tornadic storms' electrical activity caused a TV set tuned to channel 2 to glow brightly. Popular magazines ran articles explaining how to turn a TV set into a tornado monitor during a severe storm. Nowadays, the nicest thing one can say about tornado detectors is that they are unproven and probably generate numerous false alarms. And the "channel 2" fad faded out years ago, along with disco dancing and gasoline lines.

NASA launched the Optical Transient Detector (OTD) satellite on April 3, 1995. It monitors lightning flashes from orbit in space. On April 17, the OTD passed over a violent storm in Oklahoma. The instrument detected a surge in the number of lightning bolts, to more than 60 per second. About a minute later, ground observers saw a tornado touch down. NASA claimed OTD offers "the possibility of identifying the formation of tornadoes and severe storms from space."

Was the lightning surge really linked to the tornado touchdown? Or was it a coincidence? Hugh Christian, OTD principal investigator at NASA's Marshall Space Flight Center, said the satellite recorded almost 200 lightning flashes during three minutes over the storm. In contrast, the ground-based National Lightning Detection Network (NLDN), a network of sensors across the United States, detected only nine flashes from the storm. Why the difference? Because NLDN detects only flashes from the clouds to the ground, whereas OTD also sees "intracloud" lightning (lightning between clouds). Perhaps tornado formation is more closely related to intracloud lightning than cloud-to-ground flashes. Christian suggests that a strong downdraft caused the lightning and tornado simultaneously. "Further research and the experience gained with this lightning instrument could help develop sensors for real-time severe weather warnings and assist with identification of the formation of tornadoes," Christian says.

Scientists seek other ways to use satellites to distinguish between thunderstorms that will drop tornadoes and those that won't. Tall, severe thunderstorms have powerful updrafts that punch through the cloud's icy roof and into the lower part of the stratosphere. These dome-shaped protrusions are called overshooting tops. From space satellites, overshooting tops resemble warts on smooth skin. Fujita and others have explored possible links between overshooting tops and the time of a tornado touchdown. Scientists also study possible connections between tornadoes and other cloud top features, such as mysterious V-shaped cold spots.

NASA began launching a new series of weather satellites in 1994. The space agency launched GOES-8 (Geostationary

Operational Environmental Satellite) from Cape Canaveral aboard an Atlas Centaur rocket on April 13, 1994. The $220-million, 22,000-mile-high probe is "geostationary" because it stays above the same point on Earth—75 degrees west longitude, off the east coast of the United States. The new satellites can "zoom in on a significant weather event every five minutes while continuing to provide overall coverage," NOAA officials say. These "eyes in space" could prove invaluable for monitoring severe storms such as tornadoes, hurricanes, flash floods, and hail storms.

Researchers are studying ways to strengthen homes and other buildings against tornadoes. Many buildings could survive tornadoes if better built, say scientists at Texas Tech's Institute for Disaster Research (IDR). Unfortunately, structural engineers have as much trouble persuading people in Tornado Alley to spend several hundred dollars making their homes tornado-resistant as they have persuading Californians to make their homes earthquake-resistant. "It won't happen to me" is the usual excuse.

Since 1987 the insurance industry has paid a fortune for windstorm damages. In response, the industry is pressuring states to tighten building codes. It's an uphill battle: Politically powerful homebuilders associations oppose any new regulations.

At IDR, scientists simulate tornado damage by using a cannon to fire high-speed projectiles (lumber, pipes, etc.) at walls. In the 1970s, IDR scientists concluded that even the worst tornado damage could be caused by winds as weak as 275 mph. Their research has helped building engineers develop new structural engineering codes for wind-resistant buildings.

"Some structures—such as nuclear power plants—need complete protection from tornadoes regardless of the high costs of design and construction. Public facilities also need protection, but local governments rarely have the necessary funds," said architect Harold W. Harris and IDR's Kishor C. Mehta and James R. McDonald in a 1992 article for *Civil Engineering* magazine. "Yet a fairly high degree of defense is available for a minimal amount of money." They urge businesses and schools to establish enhanced tornado protective areas (ETPAs), where people can safely hide during a tornado. The ETPA should be on the same floor as its occupants: "Moving several hundred students up or down stairs is time-consuming." A building's interior may be safe even if it wasn't designed for tornadoes. For example, a Xenia, Ohio, high school was "virtually destroyed" by a tornado in 1974. Yet the first-floor hallway was undamaged.

Old buildings are especially vulnerable, particularly those with unreinforced masonry. Vulnerable masonry can be reinforced by inserting steel rods into the walls. Special frames and trusses also enhance building stability.

Facilities in tornado-probe areas may prefer to buy a dedicated tornado shelter rather than to retrofit their existing plant. One of the world's top particle physics laboratories is Fermi National Accelerator Laboratory (Fermilab) in rural Batavia, Illinois, west of Chicago. Recently, lab officials unveiled what they claim is the world's strongest above-ground tornado shelter. Shaped like a loaf of French bread, the above-ground shelter can hold several dozen people. Its walls are steel-reinforced precast concrete. North Star Chicago Precast of Naperville, Illinois, made it and markets similar shelters. Other firms sell shelters as small as in-home units for families. The house may collapse, but the shelter will (in theory) remain intact.

Steel-frame homes are an intriguing new option. An estimated 50,000 steel-frame homes were built in the United States in 1995 (about 4 percent of the total)—100 times as many as in 1992. Builders claim that well-built steel-frame homes resist high winds and earthquakes (not to mention termites!) They cost slightly more than wood-frame homes. The American Iron & Steel Institute hopes to seize 25 percent of the new-home market by 1999.

What do you do if you see a tornado approaching?

First, be aware of an important distinction:

(1) A "tornado watch" means a tornado *might* appear but hasn't been seen yet.

(2) A "tornado warning" means a tornado has been detected. Head for shelter immediately. If you are driving, head for a safe place.

During a tornado watch, turn on a radio or TV and listen for further advisories. Protect valuables—for example, move your car into the garage. Otherwise the tornado might turn it into a missile and hurl it into your home or someone else's.

If you have time, move other potential missiles outside your home (lawn furniture, lawnmowers, pink flamingoes, etc.) indoors.

If you hear sirens, they mean you should stay indoors and take cover.

However, don't wait to hear warning sirens before you head for shelter. Sirens usually *don't* wail before a tornado hits. (Remember Barneveld!)

Do NOT get into your car to flee the area! Research shows that motorists are much more likely to be killed or injured than people who stay at home. "Do not attempt to out-drive a tornado," warns the Federal Emergency Management Agency. "They are erratic and move swiftly."

Do NOT run around your home to open windows! Experts no longer believe this is a good way to protect your home. Until about 1980, emergency officials believed that buildings "exploded" from internal air pressure as the low-pressure tornado funnel passed overhead. Then researchers at the Institute for Disaster Research studied building damage in detail and concluded building damage was caused by high winds, not by explosions. Indeed, you might *increase* the risk of damage by opening windows, which admits high winds that could accumulate in the center of the house, rise, and shove the roof off! Also, if you run around opening windows, you risk being hit by flying glass and other debris.

Another myth is that the southwestern part of a building is always the safest place. There may be safer spots.

If you're in a mobile home during a tornado warning, leave and head for a safe structure or low, protected ground. The jokes about trailer parks being "magnets" for tornadoes are based on tragic truth: Trailers are too flimsy to withstand high winds.

If a tornado approaches, head for the basement and hide under a sturdy structure (such as a heavy table or workbench). If you don't have a basement, then hide in a small interior room—say, a bathroom or closet.

Avoid windows (they may shatter) and outside walls (they may collapse).

Avoid big open rooms such as gymnasiums and auditoriums. They tend to collapse. If you're in a shopping mall, find the center of the building and go to the lowest level. (Do *not* go to your parked car!) Again, avoid windows and big, open rooms.

If a tornado is near and you're driving, get out of the vehicle and lie down in a ravine, ditch, culvert, or other low area. Place your hands behind the back of your head and neck to avoid neck injury. If you're in a low area, keep an eye out for flash floods.

After the tornado has struck, check for gas leaks; know how to turn off your utilities if necessary. When you leave the building, watch out for fallen electrical lines and broken glass. It wouldn't hurt to have a fire extinguisher nearby. (Be sure that you know how to use it—and that it hasn't passed its expiration date!)

The best safety precautions are those taken long before the tornado hits. To protect yourself, your family, and your neighbors, you might consider:

—Practicing "disaster drills" with your family.

—Stocking emergency supplies in advance. These include a flashlight (along with a fresh, unopened pack of batteries), candles (with matches in a waterproof container), first-aid items (again, in a waterproof container), and a transistor radio.

—Joining a local tornado spotter's group. For details, call your local emergency preparedness office.

—Learning cardiopulmonary resuscitation (CPR). If performed correctly, CPR saves lives. I emphasize the word "correctly": Studies show that some laypeople perform CPR improperly and someone dies as a result. The best way to learn

CPR is by taking a class at a responsible agency—such as your local Red Cross, hospital, or health department.

—Obtaining a copy of *Are You Ready?*, FEMA's comprehensive guide to preparedness for every conceivable major disaster from tornadoes to earthquakes to hurricanes to hazardous materials spills.

If you live in a trailer park, find out where your trailer park's emergency shelter is located—if it has one. If it doesn't, ask the park manager to install one. Your outspokenness may end up saving more lives than your own. Make sure the shelter is accessible to people who are frail, handicapped, or need medical assistance.

For children, NOAA and FEMA offer a cartoon-filled brochure called "Owlie Skywarn's Weather Book."

Spanish-language tornado warning brochures are available from NOAA, the American Red Cross, and FEMA. The Spanish-speaking population of the United States is rising, especially in tornado-prone border states. If you have neighbors who are new to the United States and some of them are still learning English, you would be doing them a favor by bringing these brochures to their attention, or by posting them at a central location (say, a community center or place of worship).

Further advances in the understanding of tornadogenesis may lead to improved tornado forecasts and warnings and even to the remote possibility of modifying tornadoes. . . .
—Erik N. Rasmussen, Jerry M. Straka, Robert Davies-Jones, Charles A. Doswell III, Frederick H. Carr, Michael D. Eilts, and Donald R. MacGorman, writing in June 1994 *Bulletin of American Meteorological Society*

A quarter of a century ago, "tornado modification" was a hot topic. Now almost no one talks about it. Why? Will it ever make a comeback? Have we any hope of quelling the twister's killer winds? Or is its ferocity forever untamable?

Indeed, the entire science of "weather modification" has lost much of its former allure. Federal funding for research on weather modification neared $20 million per year in the early 1970s. By the 1990s it had fallen to a few million per year. In 1995, Congress appeared close to killing NOAA's weather modification program altogether.

The weather modification movement, once so strong, withered for many reasons. For one thing, it never lived up to its boldest hype. Deserts did not bloom; hurricanes were not diverted from American shores. Scientists never agreed whether cloud seeding had enough impact on rainfall, hail, or lightning to be economically attractive. Weather modification also lost support because of the emergence of the environmental movement, which changed the way that Americans perceive nature. Nowadays people are more nervous about tampering with nature; they fear we'll screw it up even more than we already have. The movement also undermined key rationales for two weather modification programs—rainfall generation and lightning suppression. Once, energy planners advocated increasing rainfall to feed water to hydroelectric dams; now hydroelectric dams are politically unpopular. Also, the U.S. Forest Service once tried to stop lightning to prevent forest fires. But now, forest experts view lightning-caused forest fires as a natural part of the ecosystem: The fires clean out old brush and allow new life to thrive.

Once, scientists hoped to use cloud seeding to quell tornadoes. They speculated that seeding would cause rain that, as it fell to Earth, would unleash a cool gust of air. The gust might kill a tornado by suffocating its updraft, they theorized. The only known attempt to do so occurred in the early 1970s. In 1972, at the American Meteorological Society meeting, T. J. Henderson and W. J. Carley reported the results of their experiment. They had seeded thunderstorms that generated six tornadoes. The twisters were unaffected.

The last serious attempt at tornado modification was made by a noted aerospace scientist, Chieh-Chein Chang.

In the early 1930s, as a youth in Nanking, China, he saw an aircraft factory that had been wrecked by a tornado. He began to wonder if there was a way to weaken tornadoes.

Later he moved to America and became a specialist in fluid dynamics, nuclear fusion, advanced aircraft, and futuristic schemes for spaceflight. While heading the space sciences department at Catholic University in Washington, D.C., he built a tornado machine in his lab, one that generated a smokey "twister" that spun 20 times per second. He injected confetti and soap bubbles into the vortex to study its air flow.

A tornado endures partly because new air keeps rushing into its low-pressure core. What would happen if one momentarily _raised_ the air pressure inside a funnel? Chang decided to find out. He filled a rubber balloon with an explosive mixture of hydrogen-oxygen gas. He floated the balloon in the middle of the tornado machine. As the mini-twister whirled, Chang pressed a button that ignited a spark; the balloon detonated in a bright flash. The gas surged through the funnel, raising its air pressure and momentarily disrupting the funnel. "Thus," Chang asserted, "we have established the mechanism of killing the tornado-like vortex in the laboratory for about one second."

What would happen if he detonated a bigger balloon inside a real tornado? He speculated that if the thundercloud were moving fast enough, then the top part of the tornado would pull away while its lower part dragged on the ground. Therefore the vortex would pull itself apart, destroying the tornado "before it can do damage to the buildings and installations downstream."

"If he's right," _Science Digest_ said in 1969, "you might see tiers of balloons anchored over cities and installations in the U.S. tornado belt, ready to be launched when a tornado begins to form."

In a 1976 lecture at a tornado conference in Lubbock, Texas, Chang suggested firing the explosives from an antiaircraft gun or other launcher "which can deliver the explosive device . . . right at the middle of the moving tornado vortex tube within the accuracy of 50 feet. . . . Where and how to find such a swift and accurate delivery system is of our future concern. As a first field test, the explosive charge should contain 200 to 500 pounds . . . of liquid hydrogen or . . . other more suitable and safer gases. . . ."

Chang decided to test his idea in the field—on dust devils, not tornadoes. He and five students traveled to White Sands Proving Ground in New Mexico, where they set up their equipment.

But bad luck dogged him just as it had dogged the futuristic experiments of Vernon Rossow and Stirling Colgate. "Storms damaged our rigs of instruments and reduced the number of occurrences of the dust devils to very few," Chang lamented. "We worked in a wrong season! . . . Our time and effort were wasted. . . . We suffered so much that we have not tried again since then, although my dream may become true someday."

Nowadays, the "dream" of tornado modification isn't quite dead, but it certainly is slumbering. Rasmussen and his colleagues mentioned it as a "remote" possibility in a 1994 paper for the *Bulletin of the American Meteorological Society*. "I just got another proposal like that across my desk," he told me in late 1995. "It's a confidential proposal, and I'm not saying who it's from, but this person is a *very* good fluid dynamicist. He thinks he sees the Achilles heel in tornadoes. He proposes a way to release a bunch of heat into the vortex—he thinks that will break it up.

"I don't know, though," Rasmussen added, frowning. "I'd be real reluctant to release a bunch of heat into the vortex, because I'm worried that would *intensify* the tornado, not break it up—intensify it by a factor of maybe two or three.

"I think we could stop a tornado if we could produce a lot more *cold* air in its vicinity. How? By cloud seeding. You'd have to inject the chemicals into the cloud from below, though. You can't drop anything from above into a tornado, because the updrafts are already rising at 60 mph." In theory, seeding would cause water vapor to condense and fall as rain. That, in turn, might create a wave of cold air that would plummet to the ground. Then, with any luck, the cold air would extinguish the tornado as a stiff breeze snuffs a burning candle.

Then Rasmussen seemed to think better of his idea and shook his head. "It's all fantasyland. I don't know *what* to do. Tornado modification would be extremely difficult anyway because the storm is so energetic and the storm environment is so difficult to work in. When we were on the ground at Friona, we were *10 kilometers* from the tornado, yet the visibility was low and power lines were blowing down. . . . *That's* the kind of environment you're dealing with."

Straka is blunter. He has a one-word reply to anyone who talks about tornado modification or control: "Bullshit."

"Do a scatter analysis of the energy," Straka scoffs. "How are you going to redistribute or dissipate all the energy associated with a tornadic storm? That's a huge amount of energy, equivalent to I don't know how many atomic bombs—lots of people have figured it out on an envelope." He laughed, recalling a letter he once received from an interested citizen who proposed a way to "bomb" tornadoes.

"We're not going to control tornadoes—not in my lifetime," Straka advises. "If we do, I'll eat crow."

Joseph Golden hasn't lost hope, though. The veteran waterspout and tornado researcher gave a speech at an American Meteorological Society meeting in early 1996 in Atlanta, where he recommended a "scheme for speeding up the natural decay of the vortex" via cloud seeding. Initial experiments could be conducted on waterspouts.

It's an imaginative idea—but the times are unfriendly to imaginative ideas. Golden's current job is head of weather modification research for NOAA. In late 1995, Congress appeared close to axing his program. For the foreseeable future, it appears, tornadoes have nothing to fear from the human race.

The sky is literally the limit for tornado chasers. One day astronauts may pursue tornadoes across the orange-red sands of Mars. Space scientists who study close-up photos of Mars have noted bright glows and mysterious dark trails on the planetary surface. The bright glows, they say, are dust devils. The trails may be tornado tracks, etched in the rust-colored soil by passing families of twisters.

The future is bright for Earth-based tornado chasers, too. VORTEX is over, but the scientists have gathered mountains of data that they'll spend years analyzing. Smaller versions of VORTEX may be held in coming years—if federal funds hold out.

New technologies may aid vortex chasers. Joseph Golden of NOAA thinks helicopters might become the choice chase vehicle of tomorrow. They're unhampered by roads and can travel at top speed. In 1993, on an expedition funded by the National Geographic Society, he flew in a helicopter past waterspouts in the Florida Keys. He also speculates about turning speedboats (with weather instruments) into chase vehicles for waterspout research.

A priority of future research should be the launching of instrumented probes into funnels, perhaps with small rockets (à la Stirling Colgate) or remotely controlled small airplanes. The plane couldn't survive a flight into the funnel, but a missile might. Maybe scientists could acquire a few military surplus "smart" missiles, similar to those used in the Gulf War, and outfit them with video cameras and weather instruments. There are worse ways to beat swords into plowshares, anyway.

One imagines what the little rocket might see as it plunges through the hurricane-force wall of the twister and, guided by remote control, soars up and down the funnel, measuring wind speeds and temperatures. Entire fleets of rockets might be launched into tornadoes, to record sights never before seen.

And who can say that someday, a human might not follow them? Only a fool would deliberately attempt to enter a tornado, of course. But foolhardiness is the consistent trait of all pioneers. Most chasers are young; and like most young people, they believe they're going to live forever. Sooner or later, against all common sense, by some means we can only dimly imagine, one of them will likely attempt the impossible—to deliberately enter a tornado.

It would be an insane act, of course. So was Lindbergh's flight across the Atlantic.

Anton Seimon is about as young now as the white-haired Robert Davies-Jones was a quarter of a century ago, when scientific chasing was born. Over that time chasers have changed our view of tornadoes. What sights will they see over the *next* twenty-five years? The interior of the tornado remains as remote, as little understood, as the surface of Pluto. Will it remain forever so? "I wouldn't be surprised," Seimon said with a faraway sound in his voice, "if in the next ten years, someone survives a trip into that no man's land with a video camera."

*A*FTERWORD

The University of Illinois's "virtual reality" thunderstorm is crude compared to the "tornadoes" of the film *Twister*. *Twister*'s tornadoes make a quantum leap forward in the techniques of "digital" moviemaking. They also foretell the future of film: Tomorrow's moviemakers will depict on screen *anything* that the mind can imagine.

Until now, the most celebrated computer-generated film images involved the simulation of objects with geometric or "continuous" surfaces. Classic examples include *Terminator 2*'s murderous linoleum floor, which rises and transforms into a cyborg who kills a cop; and the watery "pseudopod" in *The Abyss*, which emerges, like a long, translucent worm, from a sunken spaceship and slithers toward a female scientist, then changes shape to imitate her facial expressions.

But *Twister* involves a radically different type of digital effect. A tornado is not a single object with an unbroken surface. Rather, it is a constantly moving *swarm* of objects—billions of them. In nature, a tornado is a vortex of high-speed wind—moving air—that is, by itself, invisible. (You can't see air.) What makes the tornado visible is the objects that swirl around it: water droplets (the cloud-like "condensation funnel" that sheathes the vortex), dust particles, and larger chunks of debris such as cows, cars, and roofs. These objects are, in effect, the tornado's "clothes" that make it visible to the naked eye.

How could computer moviemakers simulate such a complex, high-speed maelstrom of objects? The answer is that, initially, they weren't sure they could. Before giving *Twister* the go-ahead, executive producer Steven Spielberg asked special effects experts at Industrial Light & Magic (ILM) in Marin County, California—where *Jurassic Park*'s digital dinosaurs were born—to demonstrate that they could create realistic-looking tornadoes on computer screens. The ILM wizards made a test film by putting a small movie crew in the back of a car and filming the driver as he drove toward a building. Then they digitized the images and used computer techniques to "paint" a tornado in the distance as it tore a building to bits, then hurled another vehicle at his windshield. The results were stunning—as convincing as a documentary film on tornadoes.

"*Twister* couldn't have been made without digital effects," director Jan De Bont declares.

Twister depicts a wild menagerie of tornadoes, as diverse in shape and size as real-life twisters. Some are serpentine, whitish funnels that snake elegantly over a body of water. One is a menacing, mud-brown torrent that almost kills the film's heroes, tornado chasers Jo and Bill Harding (played by Helen Hunt and Bill Paxton), then steals Jo's car. And yet another is a hellish funnel into whose interior Jo and Bill gaze at the film's climax.

The actors performed in location shooting in Oklahoma and Iowa in the first half of 1995. Then the footage was sent to ILM, where tornadoes were digitally added to the footage.

That's easier said than done. Many laypeople seem to have the impression that computer modeling is easy—"that you press a button, let the computer do it, and go home," ILM visual effects supervisor Stefen Fangmeier notes. "And it's really not like that; it's more like a Renaissance craft, like manual labor."

Visual effects producer Kim Bromley agrees. Dazzled by the film potential of digital technology, "writers are out there writing the most outrageous things you heard in your life," she says with a laugh. "They write, 'And the Death Star explodes . . .' in several keystrokes. Eight months and tens of thousands of dollars later, we've managed to turn those keystrokes into a realistic-looking sequence of the Death Star exploding."

The most famous tornado movie is, of course, *The Wizard of Oz* (1939). The film was directed by Victor Fleming and starred Judy Garland as Dorothy, the Kansas farm girl who, with her Scotch terrier Toto, travels aboard a twister to the marvelous land of Oz. *The Wizard of Oz*'s tornado—a dark, writhing funnel—is one of the unforgettable images of filmdom. It thrilled a generation of Baby Boomers who grew up watching *Oz* reruns on TV. The "tornado" is, in fact, a 35-foot-long muslin wind-sock.

Oz's tornado is still wondrous to behold. But scientifically speaking, it is—to put it politely—somewhat idealized. In reality, many tornadoes lack the *Oz* funnel's smooth, sinuous shape. Also, very few whip back and forth so elegantly, like bullwhips. Since the early 1980s, many Americans have seen close-up videotapes of real tornadoes on TV; they've learned that many twisters are, in fact, complex and highly unstable objects. Many look less like smooth funnels than ragtag assemblies of individual clouds.

Such public awareness put pressure on *Twister*'s makers because "we're having to duplicate nature," says producer Kathleen Kennedy. But there was no practical way to simulate a "modern"-looking tornado with a physical model like *Oz*'s wind-sock. Ordinary animation was out of the question, too. Some films have shown twisters that were actually animated cartoons superimposed on live-action shots; and unfortunately, they *look* like cartoons. In short, digital technology offered the only hope for creating realistic-looking tornado images that would satisfy the savviest tornado buffs.

To demonstrate digital construction of a tornado, Fangmeir switched on his Silicon Graphics computer workstation. A skeletal, cartoon-like outline of a tornado appeared on the screen. "It looks like a giant caterpillar," Fangmeir said, moving the mouse to manipulate the "tornado." It resembled a wire-mesh outline of a funnel with blue points to mark different positions. The funnel was situated on a larger wire-mesh grid that represented the terrain—specifically, a drainage ditch where Jo and Bill have fled for safety. "The animator controls the shape of the tornado and the rate at which it moves across the terrain."

Next step: to give the skeletal "tornado" a realistic-looking texture—in effect, to give it "skin." Most tornadoes' surfaces appear rough, not smooth. How can digital artists mimic that roughness without wasting months digitally "painting" each pixel on the computer screen? ILM artists create mathematical programs that automatically generate rough-looking "fractal" (fractional dimension) surfaces. By altering the data inputted in the program, they can vary the amount of roughness on an object. (Back in the 1980s, moviemakers began using fractal mathematics to create realistically "rough"-looking surfaces, such as mountainsides. A noted example is the "Genesis Planet" created for one of the *Star Trek* films.) "Fairly complex algorithms [computerized mathematical programs] allow us to do this. Of course," Fangmeir adds with a smile, "we write completely different algorithms for dinosaurs."

With further computer tinkering, they can add colors (for example, a dirt-brown shade), sprinkle flying debris around the funnel, and make it rotate. The result is an astonishingly realistic-looking tornado. To generate such images requires "gigabytes"—billions of bytes—of computer disk space, as much as one can store in dozens of small personal computers.

Digital techniques also allowed the digital artists to synchronize the tornado's motion, frame by frame, with the constant bouncing and jumping of the camera. There's a lot of "bouncing" in this movie as Jo and Bill race in their truck across the countryside or flee by foot from a twister. In old-time Hollywood, filmmakers enjoyed the luxury of a static camera: It sat still and filmed a scene while actors moved in front of it. But nowadays, bouncing camera motion (the "hand-held camera" look) is a staple of action films; audiences *expect* it. A bouncing screen image contributes to the viewers' sense of realism and replicates what their eyes would actually see—their field of view—if they were *with* Jo and Bill on a chase. So ILM digital artists couldn't simply "paint" a tornado on a screen; rather they had to "model" the entire scene three-dimensionally in a computer to ensure that the tornado retained the correct position as the camera jiggled around.

Another digital effect is "motion blur." In early science-fiction films, dinosaur puppets were animated frame by frame, which gave their motion a "jerky" look. In the real world, objects move continuously, and the faster they move, the more the eye perceives their motion as blurred. So by creating artificial "blur," digital experts enhance the realism of a moving digital object. They create the blur by using the computer to spread out the moving object's image. Motion blur was a key factor in contributing to the verisimilitude of *Jurassic Park*'s dinosaurs. (If you take a videotape of the film, stick it in your VCR, and freeze-frame it during a special effects scene, you'll see that individual components—say, one of the *Tyrannosaurus rex*'s arms—are blurred.) Likewise in *Twister,* swarms of debris (all computer generated) are motion-blurred to enhance their realism.

Fangmeir showed how ILM digital artists designed a "cow" that, in one memorable scene, flies past the window of Jo and Bill's truck. Originally the cow was a zebra used in the recent Robin Williams fantasy movie *Jumanji.* Via computer techniques, they transformed the zebra into a cow. Nowadays, the cow effect is a "rather run of the mill" effect for digital artists,

Fangmeir says. Even so, it's time-consuming; he estimated that digital artists had to work the equivalent of a few man-months to create the flying cow. The "flying cow" effect is funny and, best of all (Bromley notes), the Humane Society need not worry: No cows (or other animals) were harmed in the making of this film—only digital ones.

Science-fiction movies are littered with almost-invisible "jokes." They are the visual effects experts' equivalent of writing "Kilroy slept here" on a restroom wall. For example, during the making of the *Star Wars* trilogy, a model maker glued a tiny replica of San Francisco's pyramidal Transamerica Building on the side of a giant starship. Likewise, in *Twister*, effects experts created whimsical digital images of tornado-blown debris—such as a plastic pink flamingo from someone's lawn. "One of the modelers' nicknames is Edsel," Bromley said, "so he modeled a grille from an Edsel car and made it into a piece of debris."

In January 1996, De Bont, the director, sat in a darkened room filled with ILM staffers and examined a video playout of their latest handiwork. The video screen showed the same short scene over and over. The scene started with an image of the sky taken from a camera aboard a truck; a slender, grayish tornado lurks in the distance. Then the camera pans to the left, past a damaged tractor—one of its wheels is still spinning following its encounter with the twister. The scene ends with the anxious face of the driver, Bill Harding's fiancée, Melissa Reeves, played by Jami Gertz. Reeves is *Twister*'s link to a typical moviegoer; she portrays a nonchaser who finds herself caught in a tense tornado chase. As the scene plays over and over, De Bont murmurs his instructions to the ILM staffers: Darken this, lighten that . . . De Bont has watched mountains of videotapes and films of real-life tornadoes, and consulted with numerous scientists at the National Severe Storms Laboratory in Oklahoma; he has become a kind of tornado expert. If anyone in Hollywood knows what a movie twister should look like, it's De Bont.

Watching the video screen, you might think the tornado is

the only digital effect in the scene. It isn't; there is another one, although you'd never guess it. Remember the damaged tractor with the spinning wheel? It *looks* like a real tractor, but it isn't. It's a complete digital illusion, manufactured in the computer and added to the scene after it was filmed. De Bont came up with the idea for the tractor after shooting a scene where a tractor was blown aloft. Wouldn't it be nice, he thought, to show what happened to the tractor? By then it was too late to go back to Oklahoma and reshoot the scene, but that's no problem in the digital age. In the comfort of an ILM office, digital artist John Stillman blended a computerized image of a tractor into a digitized reel of the movie footage. As an added touch, he made its wheel spin. The "tractor" effect passes so quickly that in the final film, most viewers may miss it entirely. Yet it's one of the countless subtle touches that give the film its gritty, realistic look.

Such "invisible" digital effects are increasingly common in films. You may have heard about the digitized "feather" that floats down from the sky and lands at the protagonist's feet in *Forrest Gump*. But there are many other examples, some so mundane that—like the tractor—you'd never suspect they were digitized. A slighty-too-revealing woman's swimsuit in *My Father, the Hero* is broadened to suit family audiences. An advertising sign for Taco Bell, a company unfamiliar to non-U.S. audiences, is changed to a Pizza Hut sign for foreign releases of *Demolition Man*. Digital effects also allow a filmmaker to turn a few hundred cheering sports fans into a stadium packed with tens of thousands of people.

Naturally, some Hollywood veterans fear that digital technology will put a lot of crafts workers—lighting experts, model makers, and set designers, for example—out of work. Why build an elaborate stage set when you can digitize one in a computer? Why hire ten thousand extras when a few hundred can be "cloned" in the computer?

Twister's special physical effects boss, John Frazier, recalls that years ago, "I fought [digital effects] like everybody else and thought, 'Oh man, I've got to find another career.'" As it turned out, "what has happened is they're making *more* movies because of digital effects. *Twister* would never have been made without digital—it would still be on the shelf. So even though digital has taken some work from us, it has made *more* work than it has taken away."

De Bont says: "A lot of people are very worried about all these new technologies. But I think it's an *incredible* tool. The possibilities you have to 'paint' an image right now are *so* enhanced. In the past, the footage was totally dependent on what the [filming] situation was like. You didn't have *time* to wait until the weather was right, until it starts raining or hailing. *We* can create rain or hail whenever we want!"

Not without considerable effort, though. Digital work can be exhausting. "At times I go home and I want to cry, I'm so tired," Bromley said. "But the first time I saw a dinosaur on film, I knew filmmaking would never be the same."

To actors, digital filmmaking poses a special challenge. During the shooting of a film, they must react realistically to amazing sights—say, a tornado or a dinosaur—that aren't really there. Through the spring and summer of 1995, *Twister* leads Bill Paxton and Helen Hunt screamed and yelled and pointed at imaginary "tornadoes" in the empty Midwestern sky. The tornado images were added to the film months later by ILM's digital artists.

In one sense, Hunt and Paxton were doing what any actor does, whether it's in a blockbuster movie or a community dinner theater: simulating human emotions and behaviors in a simulated environment. Some actors, especially those of the "Method" school of acting, resort to extreme tactics to make their performances convincing. For example, Dustin Hoffman medicated himself so he'd look realistically exhausted during the torture scene in *Marathon Man* (1976); to which his exasperated costar Laurence Olivier responded: "Oh, gracious, why doesn't the dear boy just *act?!*" [See Donald Spado's *Laurence Olivier—A Biography* (1992).]

Paxton says: "I'm more of the 'Jimmy Cagney' school—when I have to do an emotional scene, I just throw myself into it." Not without preparation, however: He psyched himself for *Twister's* tensest scenes by watching a commercially available videotape, *Tornadoes: The Entity.* "It's an hour of tornado footage, and it's got this weird Philip Glass-type music throughout it. I just 'fed' all those images into my head." Paxton also participated in a tornado chase with a VORTEX team. For him, the road trip to the Texas "panhandle" was a trip down memory lane. He's a native of the Midwest and grew up in towns such as Fort Worth, Texas. "We passed by drive-in movies that I remembered taking teenage girls to. I'd say, 'Gosh, I remember taking Nan Brogan to that theater!'" He didn't see a tornado, though.

He also read numerous accounts of historic tornadoes, such as the ghastly Tri-State Tornado of 1925 that supposedly scattered human limbs around the countryside. "The Tri-State must have looked like four *Titanics* coming at you," Paxton says. "The night before I met Jan [De Bont, the director], I just crammed my head with tornado stories. I was the first and only actor he saw for the part."

As for Hunt, she psyched herself for *Twister's* climax by recalling a sight she had recently seen: the birth of a friend's baby. That stirring memory—the sight of nature delivering a new life into the world—prepared her, ironically, for the imaginary "sight" of nature at its cruelest: a tornado.

Twister is a harbinger of twenty-first-century filmmaking. What digital wonders will future moviegoers see? Will we see entirely digital worlds? Even digital actors? The Walt Disney film *Toy Story* is a sign that all-digital movies are not only technically feasible but can draw large audiences and critical acclaim.

History hints at what lies ahead. In 1968, Stanley Kubrick directed *2001: A Space Odyssey,* whose haunting special effects paved the way to the swarm of super-science-fiction films of the next two decades, notably George Lucas's *Star Wars* series. Likewise, the digital effects of mid-1990s films might lead, in the next century, to fantastic sights never before seen on the silver screen. One might say that *2001* inaugurated the age of "realistic" special effects, while 1993's *Jurassic Park* launched the age of *hyper*-realism. With the release of *Jurassic Park*, entire-screen digital images were at last indistinguishable from photographic reality. It's as if the dinosaurs were creatures in a National Geographic TV documentary. And ILM digital artists made *Jurassic Park*'s dinosaurs possible.

After *Jurassic Park* came out, film experts recognized their world was changing beyond recognition. "With computers, for the first time, you're working at the molecular level of film," said Ron Magid, who writes about special effects for *American Cinematographer* magazine.

Where will it lead? Recently Lucas, ILM's owner, has championed the concept of the "digital studio." Backdrops can be stored on computer disks, then digitally added to movie scenes whenever necessary. The result, Lucas says, could be a huge savings in money. Actors could, for instance, perform on a soundstage. Later a completely realistic-looking backdrop—say, a Hawaiian volcano or the white wasteland of Antarctica—could be digitally added to the scene. Filmmakers would be freer to pursue their cinematic visions, unconstrained by geography or finances. Conceivably, every filmmaker could imitate the global scope and majestic visual style of David Lean. (Lean, by the way, is one of De Bont's favorite directors. He also greatly admires Kubrick, and pays homage to him in *Twister*: A scene from Kubrick's *The Shining* appears on a drive-in movie screen as a tornado approaches.)

Eventually, filmmakers may routinely use computers to bring actors such as Marilyn Monroe "back to life" on movie screens. Stars might even "license" their faces to filmmakers for use long after the actor is dead and buried. Where will it all end? Will Humphrey Bogart again walk the streets of San Francisco in *Maltese Falcon 2?*

"Instead of doing one movie a year, Arnold Schwarzenegger could do ten movies a year and still have time to have a wonderful life," Magid said. (Indeed, thanks to computers, a young girl's face was transposed onto the body of a stunt woman in a particularly harrowing scene in *Jurassic Park.*) The possibilities are endless: Imagine, for example, how an actor could digitally "acquire" a new body for a nude scene.

Thanks to computers, "conceivably, you could redo the ending of *Casablanca* so Bogie would leave Casablanca on the plane, or put John Wayne in Clark Gable's role in *Gone With the Wind*," said film director Joe Dante (who made *Gremlins* and *Explorers*).

Ultimately, digital filmmaking enhances directors' ability to stir our deepest hopes and fancies. "The thing I like about visual effects is it's the art that most closely replicates your dream state," says Kim Bromley of ILM. "So you can have fairies flying around, or tornadoes in the distance, or cartoon characters come to life. . . . We can actually take these tools and make [a dream] real. And unlike a Salvador Dalí painting, you can attach sound to it and make it move."

A quarter of a century ago, the novelist John Fowles wrote:

"I saw my first film when I was six; I suppose I've seen on average—and discounting television—a film a week ever since; let's say some two and a half thousand films up to now. How can so frequently repeated an experience not have indelibly stamped itself on the *mode* of imagination? At one time I analyzed my dreams in detail; again and again I recalled purely cinematic effects . . . panning shots, close shots, tracking, jump cuts, and the like. In short, this mode of imagination is far too deep in me to eradicate—not only in me, in all my generation."

How will new film technologies affect future generations' "mode of imagination," as Fowles calls it? The power to depict any imaginable scene—no matter how eerie or unearthly—is a power that ancient peoples would have granted only to gods

. . . or to demons. Now that power is in the hands of filmmakers. The tornadoes of *Twister* are, then, more than omens of future natural disasters. Those digitized funnels also point toward new, transcendent ways of seeing—even of *experiencing* life and nature—that storytellers and novelists and poets and musicians and painters have sought for millenia, usually in vain. Now, in the same century when humanity made outer space, Antarctica, and the ocean floor part of its "backyard," it has also colonized those transcendent worlds—and all with the click of a mouse.

RECOMMENDED SOURCES

If you'd like to learn more about tornadoes and weather in general, here is a short list of useful sources. Please remember that addresses (both street addresses and World Wide Web URLs) are prone to change.

Magazines

Storm Track. Established in 1977, this lively magazine covers all the news that's news in the world of storm chasing. For subscription details, write editor Tim Marshall at 1336 Brazos Blvd., Lewisville, Texas 75067.

Weatherwise. A long-established bimonthly for weather buffs. It runs many articles on and advertisements for books and videotapes about severe weather. For subscription details, call 1-800-365-9753.

Videotapes

Numerous videotape collections of twister footage are for sale. They vary widely in quality and cost. This writer's favorite is *Tornado Video Classics I,* which includes thrilling footage of tornadoes and an excellent history of tornado research. For details, write The Tornado Project, Box 302, St. Johnsbury, VT 05819.

Books

Significant Tornadoes 1680–1991 (1993) by Thomas P. Grazulis. Advanced tornado buffs can profit from this very expensive but astonishing 1,326-page book by a tornado consultant to the Nuclear Regulatory Commission. The book describes every recorded tornado in U.S. history. It includes lengthy discussions of how tornadoes form and how scientists study them, plus numerous rare black-and-white photos. For details, write The Tornado Project, Box 302, St. Johnsbury, VT 05819.

Tornadoes (1989) by Ann Armbruster and Elizabeth A. Taylor, published by Franklin Watts. Written for middle school and high school students.

Tornadoes! (1994) by Lorraine Jean Hopping, published by Scholastic Inc. Written for grades two and three.

The Weather Book (1992) by Jack Williams. By far the best popular weather book available in the English language. Written by one of the founders of *USA Today*'s weather page, the book is rich in scientific detail and includes superb color illustrations.

Internet Sites

The Internet offers numerous weather discussion groups and World Wide Web sites.

If you have a computer and modem (ideally, at least 14,400 baud rate) you can explore the world of storm chasers at the Storm Chasers Homepage at

http://taiga.geog.niu.edu/chaser/chaser.html

The Homepage includes numerous photos of tornadoes.

The National Severe Storms Laboratory's Web site is

http://www.nssl.uoknor.edu